Multivariate Quality Control

QUALITY AND RELIABILITY

A Series Edited by

EDWARD G. SCHILLING
Coordinating Editor
Center for Quality and Applied Statistics
Rochester Institute of Technology
Rochester, New York

RICHARD S. BINGHAM, JR.
Associate Editor for
Quality Management
Consultant
Brooksville, Florida

LARRY RABINOWITZ
Associate Editor for
Statistical Methods
College of William and Mary
Williamsburg, Virginia

THOMAS WITT
Associate Editor for
Statistical Quality Control
Rochester Institute of Technology
Rochester, New York

ADDITIONAL VOLUMES IN PREPARATION

Multivariate Quality Control

THEORY AND APPLICATIONS

Camil Fuchs
Tel Aviv University
Tel Aviv, Israel

Ron S. Kenett
KPA Ltd. and
Tel Aviv University
Raanana, Israel

CRC Press
Taylor & Francis Group
Boca Raton London New York

CRC Press is an imprint of the
Taylor & Francis Group, an **informa** business

First published 1998 by Marcel Dekker, Inc.

Published 2018 by CRC Press
Taylor & Francis Group
6000 Broken Sound Parkway NW, Suite 300
Boca Raton, FL 33487-2742

First issued in paperback 2020

© 1998 by Taylor & Francis Group, LLC
CRC Press is an imprint of Taylor & Francis Group, an Informa business

No claim to original U.S. Government works

ISBN-13: 978-0-367-57932-6 (pbk)
ISBN-13: 978-0-8247-9939-7 (hbk)

**Visit the Taylor & Francis Web site at
http://www.taylorandfrancis.com**

**and the CRC Press Web site at
http://www.crcpress.com**

Library of Congress Cataloging-in-Publication

Fuchs, Camil
    Multivariate quality control: theory and applications / Camil
Fuchs, Ron S. Kenett.
    p. cm. — (Quality and reliability : 54)
Includes bibliographical references and index.
ISBN 0-8247-9939-9 (alk. paper)
    1. Quality control—Statistical methods.   2. Multivariate
analysis. I. Kenett, Ron II. Series.
TS156.F817   1998   98-2764
658.5'62'01519535—dc21   CIP

To Aliza, Amir and Efrat
—CF

To my mentors, David Cox, Sam Karlin, George Box and
Shelley Zacks who shaped my interest and understanding
of industrial statistics.
—RSK

Preface

Data are naturally multivariate and consist of measurements on various characteristics for each observational unit. An early example of multivariate data is provided by Noah's Ark for which the requirements were: *The length of the ark shall be three hundred cubits, the breadth of it fifty cubits, and the height of it thirty cubits* (Genesis 6:15). The ark was a single specimen produced by a craftsman, and statistical quality control procedures were unlikely to have been applied for ensuring the conformance to standards of the trivariate data for that product. However, modern industrial processes with mass production are inconceivable without the use of proper statistical quality control procedures.

Although multivariate methods are better suited for multivariate data, their use is not common practice in industrial settings. The implementation of multivariate quality control involves more complex calculations, creating a barrier to practitioners who typically prefer the simpler and less effective univariate methods. But this barrier can and should be removed. Modern computers have created new opportunities for data collection and data analysis. Data are now collected *en masse* and stored in appropriate databases. Adequate statistical tools that transform data into information and knowledge are needed to help the practitioner to take advantage of the new technologies. *Multivariate Quality Control: Theory and Applications* presents such tools in the context of industrial and organizational processes.

The main objective of the book is to provide a practical introduction to multivariate quality control by focusing on typical quality control problem formulations and by relying on case studies for illustrating the tools. The book is aimed at practitioners and students alike and can be used in regular university courses, in industrial workshops, and as a reference. The typical audience for industrial workshops based on the book includes quality and process engineers, industrial statisticians, quality and production managers, industrial engineers, and technology managers. Advanced undergraduates, graduates, and postgraduate students will also be able to use the book for self-instruction in multivariate quality control and multivariate data analysis. One-semester courses such as those on statistical process control or industrial statistics can rely on the book as a textbook.

Our objective was to make the text comprehensive and modern in terms of the techniques covered, correct in the presentation of their mathematical properties, and practical with useful guidelines and detailed procedures for the implemention of the multivariate quality control techniques. Special attention was devoted to graphical procedures that can enhance the book's applicability in the industrial environment.

The book presents macros written in the statistical language MINITAB™, which can be translated to other languages such as S-Plus, SAS, SPSS or STATGRAPHICS. Access to MINITAB 11.0 or higher is advisable for running the macros attached to the text. Earlier versions of MINITAB, in which the calculations are performed with single precision, are inadequate for the calculations involved in multivariate quality control.

The idea of writing a book on multivariate quality control has matured after several years of collaboration between the authors. The structure of the book reflects our understanding of how multivariate quality control should be presented and practiced. The book includes nonstandard solutions to real-life problems such as the use of tolerance regions, T^2-decomposition and MP-charts. There is no doubt that with the expansion of automatic data collection and analysis, new problems will arise and new techniques will be proposed for their solution. We hope this book contributes to bridging the gap between theory and practice and expanding the implementation of multivariate quality control.

This book distills years of experience of many people. They include engineers and operators from the metal, electronic, microelectronic, plastic, paper, medical, and food industries who were willing to pilot the implementation of multivariate quality control. We thank them all for their willingness to try new procedures that in many cases proved effective beyond expectations. Special thanks to Mr. Ori Sharon for his help in writing some of the MINITAB macros, to Mr. Brian Vogel for running the star plots, and to Mr. Russell Dekker for ongoing editorial support and continuous encouragement.

Camil Fuchs
Ron S. Kenett

Contents

ix

Multivariate Quality Control

1

Quality Control with Multivariate Data

Objectives:

The chapter introduces the basic issues of multivariate quality control. The construction of multivariate control charts is illustrated using the data from Case Study 1 on physical dimensions of an aluminum part. Several approaches to multivariate quality control are discussed and the four case studies used in the book are briefly described.

Key Concepts
- Management styles
- Quality control
- Multivariate data collection
- Multivariate scatterplot matrix
- The Hotelling T^2-distance
- The T^2-chart

Organizations can be managed in different ways. The reactive "laissez faire" approach relies on consumer complaints, liability suits, shrinking market share or financial losses to trigger efforts to remedy current problems. In this mode of thinking, there are no dedicated efforts to anticipate problems or improve current conditions. An alternative approach, that requires heavy investments in screening activities, relies on inspection of the product or service provided by the organization's customers. One set of statistical tools that apply to such a screening, or inspection-based management style, is acceptance sampling. Using such tools enables decision makers to determine what action to take on a batch of products. Decisions based on samples, rather than on 100% inspection, are usually more expedient and cost effective. Such decisions, however, only apply to the specific population of products from which the sample was drawn. There are typically no lessons learned by inspection-based management and nothing is done to improve the conditions of future products. Quality control requires a different style of management. This third management approach focuses on the processes that make the products. Data collected is used to track process performance over time, identify the presence of assignable causes and expose common causes that affect the process in an ongoing way. Assignable causes occur at specific points in time and usually require immediate corrective action, while common causes are opportunities for improvements with long-term payoffs. Feedback loops are the basic structure for process-focused management. For a comprehensive treatment of industrial statistics in the context of three management styles: inspection based, process focused quality control, and proactive quality by design, see Kenett and Zacks (1998).

Quality control is based on data sequentially collected, displayed and analyzed. Most data are naturally multivariate. Physical dimensions of parts can be measured at several locations and various parameters of systems are typically derived simultaneously. In practice, one will frequently use a critical dimension as the one to record and track for purposes of quality control. An alternative approach to handle multivariate data is to aggregate several properties and transform data collected by several measurement devices into attribute data. Under this approach, a unit is labeled as "pass" or "fail." Tracking the percentage of defective products or nonconformities is indeed a popular approach to statistical quality control. However, the transformation of continuous quantitative data into qualitative attribute data leads to a loss of information. This reduces the ability to identify the occurrence of assignable causes and impairs the understanding of the process capability. In this book we present tools that provide effective and efficient analysis of multivariate data without such loss of information.

Modern computers and data collection devices allow for much data to be gathered automatically. Appropriate tools that transform data into information and knowledge are required for turning investments in automation into actual improvements in the quality of products and processes. The main objective of this book is to provide a practical introduction to such tools while demonstrating their application. Our approach is to de-emphasize theory and use case studies to illustrate how statistical findings can be interpreted in real life situations. We base the computations on the MINITAB^{TM} version 11.0 computer package and present in some instances the code for specific computations. Previous versions of MINITAB did not have the double precision capability and caused numerical errors in several of the computations used in this book. The use of earlier versions of MINITAB is therefore not recommended for these kind of multivariate computations.

Case Studies 1 to 4 are analyzed in Chapters 3 to 11. They provide examples of statistical quality control and process capability studies. The data of Case Study 1 consist of six physical dimension measurements of aluminum pins carved from aluminum blocks by a numerically controlled machine. Case Study 2 is a complex cylindrical part with 17 variables measured on each part. Case Study 3 comes from the microelectronics industry were ceramic substrates are used with very stringent requirements on critical dimensions. Case Study 4 consists of chemical properties of fruit juices that are tested to monitor possible adulteration in natural fruit juices.

We only deal indirectly with issues related to the complexities of multivariate data collection. In some cases major difficulties in collecting multivariate data are an impediment to statistical quality control initiatives (D. Marquardt, personal communication). For example, a sample can be subjected to a variety of tests performed on various instruments located in several laboratories and requiring different lengths of time to complete. Under such conditions, collecting the data in a coherent data base becomes a major achievement.

Univariate quality control typically relies on control charts that track positions of new observations relative to control limits. A significantly high deviation from the process average signals an assignable cause affecting the process. Multivariate data are much more informative than a collection of one dimensional variables. Simultaneously accounting for variation in several variables requires both an overall measure of departure of the observation from the targets as well as an assessment of the data correlation structure. A first assessment of the relationships among the variables in the multivariate data is given by the analysis of their pairwise association, which can be evaluated graphically by displaying an array of scatterplots. Figure 1.1 presents an illustration of such a scatterplot matrix, computed for the six variables from Case Study 1 mentioned above.

The data display a clear pattern of association among the variables. For example, we observe in Figure 1.1 that there is a close relationship between **Diameter1** and **Diameter2** but a rather weak association be-

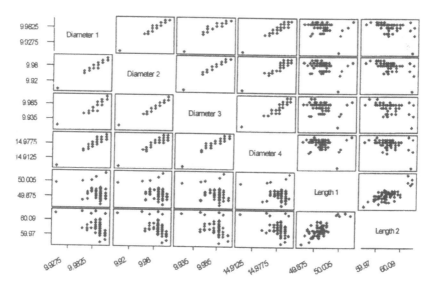

Figure 1.1: Scatterplot matrix for the 6 variables from Case Study 1 (ungrouped).

tween **Diameter1** and **Length1**. A significantly large distance of a p-dimensional observation, **X** from a target **m** will signal that an assignable cause is affecting the process. The pattern of association among the variables, as measured by their covariance matrix has to be accounted for when an overall measure of departure from targets is to be calculated. A measure of distance that takes into account the covariance structure was proposed by Harold Hotelling in 1931. It is called Hotelling's T^2 in honor of its developer.

Geometrically we can view T^2 as proportional to the squared distance of a multivariate observation from the target where equidistant points form ellipsoids surrounding the target. The higher the T^2-value, the more distant is the observation from the target.

The target vector **m** can be derived either from a previously collected reference sample from the tested sample itself (i.e., as an internal target), or from externally set nominal values (i.e., as an external target). The T^2-distance is presented in Chapter 2 and some of its theoretical properties are discussed there. The T^2 distances lend themselves to graphical display and the T^2-chart is the classical and most common among the multivariate control charts (Eisenhart et al., 1947; Alt, 1984). It should be noted that the chart, like T^2 itself, does not indicate the nature of the problem, but rather that one exists. Figure 1.2 presents the T^2-chart for Case Study 1. The vertical axis indicates the position of the 99.7-th and 95-th percentiles of T^2. We shall expand on issues related to that graph in Chapters 3 to 5 that discuss, respectively, targets that are derived externally, internally and from a reference sample. Multivariate control charts are further discussed in Chapter 6.

The multivariate analysis assesses the overall departure of the p-variable observations from their target. However, by themselves, the analysis of the T^2-statistics and of the appropriate control chart do not provide an answer to the important practical question of the detection of the variables that caused the out-of-control signal. Chapter 7 presents some approaches and solutions to this issue.

We present in Chapter 8 the statistical tolerance region approach to quality control and its related concept of natural process region. This approach is different from the classical Shewhart approach that relies primarily on three standard deviations control limits to detect shifts in the mean or the variance of a distribution. Statistical tolerance regions

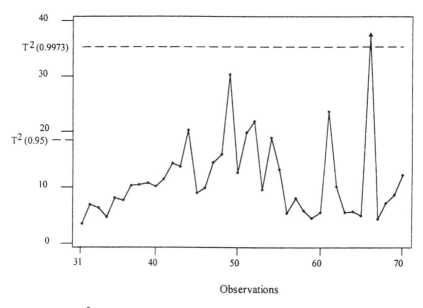

Figure 1.2: T^2-chart for the 6 variables from Case Study 1 (ungrouped).

consist of a range of values of the measured variables spanning a given percentage of the population corresponding to a process in control. This statement is made with a specific level of confidence. For example, one can construct an interval of values for which we can state, with a confidence of 0.95, that this range includes 90% of the population from a process under control. Such regions permit identification of shifts in distributions without the limitation of singling out shifts in means or variances as in the classical Shewhart control charts.

In some cases, the analyzed data are presented in batches and the quality control procedure should take this feature into consideration. Methods of analysis for this situation are presented in Chapter 9 and applied to the data of Case Study 3.

Another approach to multivariate quality control is to transform a p-dimensional set of data into a lower dimensional set of data by identifying meaningful weighted linear combinations of the p-dimensions. Those new variables are called principal components. In Chapter 10 we present the principal components approach and illustrate it using Case Study 1.

Both the T^2 distances and the principal components lend themselves to graphical displays. In Chapter 11 we compare two additional graphical displays that isolate changes in the different individual dimensions. Modern statistical software provide many options for displaying multivariate data. Most computerized graphical techniques are appropriate for exploratory data analysis and can greatly contribute to process capability studies. However, ongoing quality control requires graphical displays that possess several properties such as those presented and illustrated in Chapter 11 with the data from Case Study 1. For completeness we also review in Chapter 11 modern dynamic graphical displays and illustrate them with the data from Case Study 3.

We conclude the text with some guidelines for implementing multivariate quality control. Chapter 12 provides specific milestones that can be used to generate practical implementation plans. The Appendices contain a description of the case studies and their data and references (Appendix 2), a brief review of matrix algebra with applications to multivariate quality control, using the MINITAB computer package (Appendix 3), and a set of 30 MINITAB macros for the computation of statistics defined throughout the book (Appendix 1). The macros provide an extensive set of tools for numerical computations and graphical analysis in multivariate quality control. These macros complement the analytical tools presented in the book and support their practical implementation.

2

The Multivariate Normal Distribution in Quality Control

Objectives:

The chapter reviews assumptions and inferential methods in the multivariate normal case. Results that provide the theoretical foundations for multivariate quality control procedures are stated and demonstrated. Simulations are used to illustrate the key theoretical properties of the bivariate normal distribution.

Key Concepts
- The multivariate normal distribution
- Maximum Likelihood Estimators
- Distributions of multivariate statistics
- Base sample
- Data transformations

A multivariate observation consists of simultaneous measurement of several attributes. For example, a part coming out of a numerically controlled metal cutting lathe can be measured on several critical dimensions. Multivariate observations are conveniently represented by vectors with p entries corresponding, to the p variables measured on each observation unit. This chapter discusses inferential methods applicable to multivariate normal data. In most of the chapter we assume multivariate normality of the data and present methods which enable us to produce estimates of population parameters and construct tests of hypotheses about multivariate population parameters using a simple random sample of multivariate observations. The analysis of multivariate data poses a challenge beyond the naive analysis of each dimension separately. One aspect of this challenge stems from the simultaneous consideration of several probability statements, such as p-tests of hypothesis. This creates a problem of multiple comparisons that requires adjustments of univariate significance levels so as to attain a meaningful overall significance level. A second component of the multivariate data analysis challenge is to account for the internal correlation structure among the p-dimensions. This again affects overall significance levels.

We begin by describing the multivariate normal distribution, proceed to cover the distribution of various test statistics, and then briefly mention approaches to induce normality into non-normal data. The chapter provides a notation and a theoretical foundation for the following chapters.

The Multivariate Normal Distribution.

A p-dimensional vector of random variables $\mathbf{X} = (X^{(1)}, \ldots, X^{(p)})$; $-\infty < X^{(\ell)} < \infty$, $\ell = 1, \ldots, p$, is said to have a multivariate normal distribution if its density function, $f(\mathbf{X})$ is of the form:

$$f(\mathbf{X}) = f(X^{(1)}, \ldots, X^{(p)}) = (2\pi)^{-\frac{p}{2}} |\Sigma|^{-\frac{1}{2}}$$
$$\exp\left\{ -\frac{1}{2}(\mathbf{X} - \mu)' \Sigma^{-1} (\mathbf{X} - \mu) \right\}$$

where $\mu = (\mu^{(1)}, \ldots, \mu^{(p)})$ is the vector of expected values $\mu^{(\ell)} = E(X^{(\ell)})$, $\ell = 1, \ldots, p$ and $\Sigma = [(\sigma_{\ell u})]$ $\ell, u = 1, \ldots, p$, is the variance-covariance matrix of $(X^{(1)}, \ldots, X^{(p)})$, $\sigma_{\ell u} = \text{cov}(X^{(\ell)}, X^{(u)})$ and $\sigma_{\ell \ell} = \sigma_\ell^2$.

We shall indicate that the density of $\mathbf{X}$ is of this form by

$$\mathbf{X} \sim N_p(\boldsymbol{\mu}, \Sigma) ,$$

where $N_p(\cdot, \cdot)$ denotes a p-dimensional normal distribution with the respective parameters of location and dispersion.

When $p = 1$, the one dimensional vector $\mathbf{X} = (X^{(1)})'$ has a normal distribution with mean $\mu^{(1)}$ and variance σ_1^2, i.e.

$$f(\mathbf{X}) = \frac{1}{\sigma_1 \sqrt{2\pi}} e^{-(X^{(1)} - \mu^{(1)})^2 / 2\sigma_1^2} \qquad -\infty < X_1 < \infty$$

or

$$\mathbf{X} \sim N_1(\mu^{(1)}, \sigma_1^2)$$

When $p = 2$, $\mathbf{X} = (X^{(1)}, X^{(2)})'$ has a bivariate normal distribution with a two-dimensional vector of means $\boldsymbol{\mu} = (\mu^{(1)}, \mu^{(2)})$ and a covariance matrix $\Sigma = \begin{bmatrix} \sigma_1^2 & \sigma_{12} \\ \sigma_{21} & \sigma_2^2 \end{bmatrix}$ where σ_1^2 and σ_2^2 are the variances of $X^{(1)}$ and $X^{(2)}$, respectively, and $\sigma_{12} = \sigma_{21}$ is the covariance between $X^{(1)}$ and $X^{(2)}$. Let $\rho = \frac{\sigma_{12}}{\sigma_1 \sigma_2}$ be the correlation between $X^{(1)}$ and $X^{(2)}$.

If $X^{(1)}$ and $X^{(2)}$ are independent ($\rho = 0$) their joint bivariate normal distribution is the product of two univariate normal distributions, i.e.

$$f_{\rho=0}(X^{(1)}, X^{(2)}) = \frac{1}{2\pi \sigma_1 \sigma_2} \exp\left(-\frac{(X^{(1)} - \mu^{(1)})^2}{2\sigma_1^2} - \frac{(X^{(2)} - \mu^{(2)})^2}{2\sigma_2^2} \right)$$

In the general case ($-1 \le \rho \le 1$), the bivariate normal distribution is given by

$$f_\rho(X^{(1)}, X^{(2)}) = \frac{1}{2\pi \sigma_1 \sigma_2 \sqrt{1-\rho^2}} \exp\left\{ -\frac{1}{2(1-\rho^2)} \left(\left(\frac{X^{(1)} - \mu^{(1)}}{\sigma_1}\right)^2 \right. \right.$$
$$\left. \left. -2\rho \frac{(X^{(1)} - \mu^{(1)})(X^{(2)} - \mu^{(2)})}{\sigma_1 \sigma_2} + \left(\frac{X^{(2)} - \mu^{(2)}}{\sigma_2}\right)^2 \right) \right\}$$

We can define a set of values which includes a specific proportion of multivariate distribution. Such a set of values forms a natural process

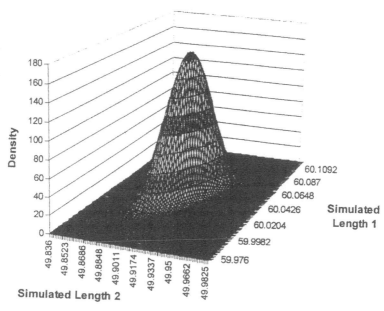

Figure 2.1a: Bivariate normal density with parameters of **Length 1** and **Length 2** from Case Study 1.

region. The values contained in the natural process region are characterized by the fact that their distances from the mean vector μ do not exceed the critical values which define the specific proportion of the population. The classical 3-sigma rule is equivalent, in this context, to testing whether an observation is contained in the central 99.73% of the reference population ($P = .9973$). In the univariate case, the respective critical values are called natural process limits (e.g. ASQC, 1983).

Figures 2.1 and 2.2 presents a bivariate normal distribution whose means, variances and correlation coefficient are equal to those found empirically in the bivariate distribution of **Length1** and **Length2** in the first 30 observations of Case Study 1, where $\mu^{(1)} = 49.91, \mu^{(2)} = 60.05$, $\sigma_1 = 0.037$, $\sigma_2 = 0.037$, $\rho = 0.723$. Figure 2.1 is a 3-D representation while Figure 2.2 displays contour plots for several values of f(Length1, Length2).

Whenever the parameters of the distribution are unknown and have to be estimated from the sample, as is usually the case, we cannot stipu-

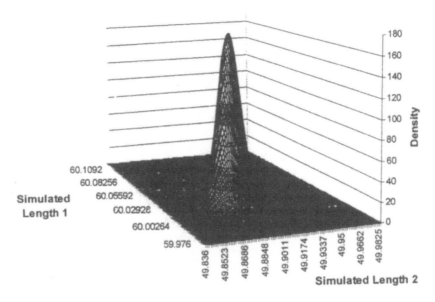

Figure 2.1b: Bivariate normal density (rotated).

late with certainty that the natural process region will contain the stated proportion of the population. Since we now have an additional source of uncertainty, any statement concerning the proportion of the population in a particular region can only be made with a certain level of confidence. Each combination of a stated proportion and a given confidence level define a region. The newly formed statistical tolerance region is used in this case instead of the natural process region. In the univariate case the natural process limits are replaced by statistical tolerance limits. The statistical tolerance regions approach to multivariate quality control is illustrated in Chapter 8.

The shape of the multivariate distribution depends on the mean vector μ and covariance matrix Σ. When one or both of these parameters are unknown, as is usually the case, they have to be estimated empirically from the data. Let $\mathbf{X}_1, \ldots, \mathbf{X}_n$ be n p-dimensional vectors of observations sampled independently from $N_p(\mu, \Sigma)$ and $p \leq n - 1$. The observed mean vector $\overline{\mathbf{X}}$ and the sample covariance matrix S are given by

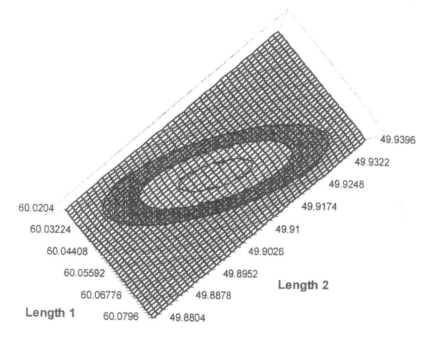

49.9396

49.9322

49.9248

49.9174

49.91

49.9026

49.8952

Length 2

49.8878

49.8804

60.0204

60.03224

60.04408

60.05592

60.06776

Length 1   60.0796

Figure 2.2: Contour plot of bivariate normal density with parameters of **Length 1** and **Length 2** from Case Study 1.

$$\overline{\mathbf{X}} = \sum_{i=1}^{n} \mathbf{X}_i / n \,,$$

$$S = \sum_{i=1}^{n} (\mathbf{X}_i - \overline{\mathbf{X}})(\mathbf{X}_i - \overline{\mathbf{X}})' / (n-1) \,,$$

are unbiased estimates of μ and Σ respectively.

We note that under the normal distribution, the observed mean vector $\overline{\mathbf{X}}$ is also the maximum likelihood estimator (MLE) of μ, while the MLE for Σ, is $\frac{n-1}{n} S$.

The (ℓ, ℓ')-th element of the S-matrix is the estimated covariance between the variables ℓ and ℓ', $\ell, \ell' = 1, 2, \ldots, p$, i.e.

$$s_{\ell\ell'} = \sum_{i=1}^{n} \left(X_i^{(\ell)} - \overline{X}^{(\ell)} \right)\left(X_i^{(\ell')} - \overline{X}^{(\ell')} \right) \Big/ (n-1) \,.$$

The diagonal elements of S are the corresponding sample variances. When a covariance is standardized by the appropriate standard deviations, the resulting estimated value of the correlation coefficient assesses the association between the two variables.

If the sample is composed of k subgroups, each of size n, and if the mean and the covariance matrix in the j-th subgroup are $\overline{X}_j$ and S_j, respectively, $j = 1, 2, \ldots, k$, then we can define the grand mean $\overline{\overline{X}}$ and the pooled covariance matrix as follows:

$$\overline{\overline{X}} = \sum_{j=1}^{k} \overline{X}_j / k = \sum_{j=1}^{k} \sum_{i=1}^{n} X_{ij} / kn$$

and

$$S_p = \Sigma(n-1)S_j / [k(n-1)] .$$

We illustrate the computations with two simulated data sets. The first sample was generated under a bivariate normal distribution with ungrouped observations. The observations in the second data set have four variables and are grouped in pairs, i.e. subgroups of size two. The advantage of the use of simulated data for illustrating the theory is that the underlying distribution with its parameters are known and we can thus observe the performance of test statistics under controlled situations. In the first data set, the parameters of the first 50 observations were identical with the values calculated empirically for **Length 1** and **Length 2** in the first 30 observations of Case Study 1, i.e.

$$\mu = \mu_0 = \begin{bmatrix} \mu_0^{(1)} \\ \mu_0^{(2)} \end{bmatrix} = \begin{bmatrix} 49.91 \\ 60.05 \end{bmatrix} .$$

and $\sigma_1 = 0.037, \sigma_2 = 0.037, \rho = 0.723$. The population covariance matrix $10, 0000 \cdot \Sigma$ is thus

$$\Sigma = \begin{bmatrix} 13.69 & 9.90 \\ 9.90 & 13.69 \end{bmatrix} .$$

Those first 50 observations simulate an in-control "base sample" (or in-control "process capability study"). The sample mean for those observations was

$$\overline{X} = \begin{bmatrix} 49.9026 \\ 60.0441 \end{bmatrix}$$

and the elements of the S-matrix are given by

$$s_{11} = \sum_{i=1}^{50}(X_i^{(1)} - 49.9026)^2/49 = .00122$$

$$s_{12} = s_{21} = \sum_{i=1}^{50}(X_i^{(1)} - 49.9026)(X_i^{(2)} - 60.0441)/49 = .00114$$

$$s_{22} = \sum_{i=1}^{50}(X_i^{(2)} - 60.0441)^2/49 = .001367 ,$$

i.e.

$$S = \begin{bmatrix} .001204 & .001123 \\ .001123 & .001355 \end{bmatrix} .$$

Appendix 3 contains a brief review of matrix algebra with applications to multivariate quality control, as well as a review of the computations which can be performed using the MINITAB computer package. If the data are in columns 1 and 2 (or, in MINITAB notation, in C1 and C2), the MINITAB command which leads to the computation of the S-matrix is simply COVARIANCE C1-C2, and we obtain the S-matrix presented above.

We mention that throughout the book the computations can be performed both with MINITAB as well as with other statistical software. Since versions of MINITAB before version 11.0 store the data with only six significant digits, the precision is lower than that achieved by software which uses double precision. The results of the computations presented in the book are those obtained by MINITAB version 11.0.

From the entries in the S-matrix sample correlation coefficient between the variables is given by

$$r = \frac{.001140}{\sqrt{.001220 \cdot .001367}} = .883 .$$

For the second data set, the parameters for the first 50 subgroups were identical with the values calculated empirically for the four diameters i.e. **Diameter 1** - **Diameter 4** in the first 30 observations of Case Study 1, i.e.

$$\mu = \mu_0 = \begin{bmatrix} \mu^{(1)} \\ \mu^{(2)} \\ \mu^{(3)} \\ \mu^{(4)} \end{bmatrix} = \begin{bmatrix} 9.9863 \\ 9.9787 \\ 9.9743 \\ 14.9763 \end{bmatrix}$$

and the covariance matrix $10,000 \cdot \Sigma$ is

$$\Sigma = \begin{bmatrix} 1.8264 & 1.7080 & 1.8195 & 1.8264 \\ 1.7080 & 1.8437 & 1.8529 & 1.8460 \\ 1.8195 & 1.8529 & 2.1161 & 1.9575 \\ 1.8264 & 1.8460 & 1.9575 & 2.3092 \end{bmatrix}.$$

The underlying distribution of the simulated data was multivariate normal with the above mentioned parameters. The data were grouped in 50 subgroups of size 2. Thus, the first group had the observations

$$\mathbf{X}'_1 = \begin{bmatrix} 9.9976 & 9.9830 & 9.9804 & 14.9848 \end{bmatrix}$$
$$\mathbf{X}'_2 = \begin{bmatrix} 9.9553 & 9.9574 & 9.9543 & 14.9492 \end{bmatrix}$$

Its mean vector of the first subgroup is

$$\overline{\mathbf{X}}'_1 = \begin{bmatrix} 9.9765 & 9.9702 & 9.9673 & 14.9670 \end{bmatrix}$$

and the covariance matrix within the subgroup ($\times 10,000$) is

$$S_1 = \begin{bmatrix} 8.9558 & 5.4181 & 5.5307 & 7.5527 \\ 5.4181 & 3.2779 & 3.3460 & 4.5592 \\ 5.5307 & 3.3460 & 3.4156 & 4.6642 \\ 7.5527 & 4.5692 & 4.6642 & 6.3693 \end{bmatrix}$$

The computations are performed similarly for the 50 subgroups and we obtained

$$\overline{\overline{\mathbf{X}}} = \sum_{j=1}^{50} \overline{\mathbf{X}}_j / 50 = \begin{bmatrix} 9.9864, & 9.9793, & 9.9752, & 14.9768 \end{bmatrix}$$

and

$$10,000 \cdot S_p = 10,000 \cdot \sum_{j=1}^{50} (50-1)S_j/[50 \cdot (2-1)] =$$

$$= \begin{bmatrix} 1.2151 & 1.1428 & 1.1600 & 1.2669 \\ 1.1428 & 1.3559 & 1.3288 & 1.3033 \\ 1.1600 & 1.2854 & 1.5728 & 1.4183 \\ 1.2669 & 1.2792 & 1.4183 & 1.6854 \end{bmatrix}$$

A MINITAB program which computes the S_p-matrix can be easily designed if we define a macro which computes for each group its $(n-1)$ S_j-matrix and sums them up. In the macro presented in Appendix 1.1, the number of parameters (p) is stored in K2 (in our case K2 = 2). The group size (n) is stored in K3 and the number of groups (k) is stored in K4. The indices of the observations in the current group are from K5 to K6. The macro (named POOLED.MTB) is given in Appendix 1.1.

The program POOLED.MTB, can be run by typing

MTB> EXEC 'POOLED.MTB' 1.

The theoretical properties of the estimates $\overline{X}$, $\overline{\overline{X}}$, S and S_p, have been extensively studied and were determined to be optimal under a variety of criteria (sufficiency, consistency, and completeness).

The distribution properties of those estimates for a random normal sample $X_1, \ldots, X_n$ are given by the following theorems (e.g. Seber, 1984):

(i) $\overline{X} \sim N_p(\mu, \frac{1}{n}\Sigma)$

(ii) If $\overline{X}$ is distributed as in (i) above, then

$$n(\overline{X} - \mu)'\Sigma^{-1}(\overline{X} - \mu) \sim \chi_p^2$$

where χ_p^2 is the Chi-square distribution with p degrees of freedom.

(iii) $(n-1)S \sim W_p(n-1, \Sigma)$

where $W_p(\cdot, \cdot)$ stands for the Wishart distribution which is considered to be the multivariate analogue of the Chi-square distribution.

(iv) If Z and D are independent random variables distributed respectively as

$$\mathbf{Z} \sim N_p(0, \Sigma_z),$$
$$f\mathbf{D} \sim W_p(f, \Sigma_z)$$

then the quadratic form

$$T^2 = \mathbf{Z}'\mathbf{D}^{-1}\mathbf{Z}$$

is distributed as

$$T^2 \sim \frac{fp}{f - p + 1} F_{p,f-p+1}$$

where $F_{p,f-p+1}$ is the Fisher F-distribution with $(p,\ f-p+1)$ degrees of freedom. If the expected value of $\mathbf{Z}$, denoted by μ_z, differs from zero, then T^2 (multiplied by the appropriate constant) has a non-central F-distribution with parameter of centrality $\mu_z' \Sigma_z^{-1} \mu_z$.

(v) Let $\mathbf{Z} \sim N_p(0, \Sigma_z)$ and $f\mathbf{D} \sim W_p(f, \Sigma_z)$, $f > p$ where $f\mathbf{D}$ can be decomposed as $f\mathbf{D} = (f - 1)\mathbf{D}_1 + \mathbf{Z}\mathbf{Z}'$, where $(f - 1)\mathbf{D}_1 \sim W_p(f - 1, \Sigma)$, and $\mathbf{Z}$ is independent of $\mathbf{D}_1$. Then the quadratic form

$$T^2 = \mathbf{Z}'\mathbf{D}^{-1}\mathbf{Z}$$

is distributed as

$$T^2 \sim fB(p, f - p)$$

where $B(p,\ f - p)$ is the central Beta distribution with $(f,\ f - p)$ degrees of freedom, respectively.

If the expected value of $\mathbf{Z}$, μ_z differs from zero, then T^2 has a non-central Beta distribution with parameter of non-centrality $\mu_z' \Sigma_z^{-1} \mu_z$.

Note that unlike (iv) above, we do not assume that $\mathbf{Z}$ is independent of $\mathbf{D}$, but rather that it is independent of $\mathbf{D}_1$.

(vi) If the sample is composed of k subgroups of size n all originated from the distribution defined above with subgroup means $\overline{\mathbf{X}}_j$, $j = 1, \ldots, k$ and grand mean denoted by $\overline{\overline{\mathbf{X}}}$, i.e.

$$\overline{\overline{\mathbf{X}}} = \sum_{j=1}^{k} \overline{\mathbf{X}}_j / k = \sum_{j=1}^{k} \sum_{i=1}^{n} \mathbf{X}_{ij} / kn, \quad \text{then}$$

$$\sqrt{\frac{kn}{k-1}}\left(\overline{\mathbf{X}}_j - \overline{\overline{\mathbf{X}}}\right) \sim N_p(\mathbf{0}, \boldsymbol{\Sigma})$$

(vii) Under the conditions of (ii) above, if $\mathbf{Y}_1, \ldots \mathbf{Y}_n$ are an extra sub-group from the same distribution, then

$$\sqrt{\frac{kn}{k+1}}\left(\overline{\mathbf{Y}} - \overline{\overline{\mathbf{X}}}\right) \sim N_p(\mathbf{0}, \boldsymbol{\Sigma})$$

(viii) If the sample is composed of k subgroups of n identically distributed multivariate normal observations, and if S_j is the sample covariance matrix from the j-th subgroup, $j = 1, \ldots, k$, then

$$\boldsymbol{\Sigma}(n-1)S_j \sim W_p(k(n-1), \boldsymbol{\Sigma}).$$

Note that this additive property of the sum of independent Wishart variables, is an extension of that property which holds for univariate Chi-square variables.

The distributional properties of $\overline{\mathbf{X}}$, S and T^2 provide the theoretical basis for deriving the distributions of the statistics used in the multivariate/equality control procedures discussed in subsequent chapters.

Those statistics assess the overall distance of a p-dimensional vector of observed means from the target values $\mathbf{m}' = \left(m^{(1)}, m^{(2)}, \ldots, m^{(p)}\right)$. Let $Y_i^{(\ell)}$, $i = 1, \ldots, n$; $\ell = 1, \ldots, p$ be n multivariate measurements and let $\overline{\mathbf{Y}}' = \left(\overline{Y}_{(1)}, \ldots, \overline{Y}(p)\right)$ be observed means.

Up to a constant, the appropriate Hotelling T^2 statistic is computed by multiplying $\left(\overline{\mathbf{Y}} - \mathbf{m}\right)'$ i.e. the transposed vector of deviations between $\overline{\mathbf{Y}}$ and $\mathbf{m}$, by the inverse of a certain empirical covariance matrix (S^{-1}) and then by the vector of derivations $\left(\overline{\mathbf{Y}} - \mathbf{m}\right)$. The Hotelling T^2-statistic has already been mentioned in the previous chapter where we presented the chart of the values obtained from the first 30 observations from Case Study 1.

For the case when a sample $\mathbf{Y}_1, \ldots, \mathbf{Y}_n$ of n independent normal p-dimensional observations is used for assessing the distance between the mean $\overline{\mathbf{Y}}$ and the expected values μ, the Hotelling T^2-statistic, denoted by T_M^2, is given by

$$T_M^2 = n\left(\overline{\mathbf{Y}} - \mu\right)' S^{-1}\left(\overline{\mathbf{Y}} - \mu\right).$$

The T_M^2 statistic can be expressed in terms of the means of the measurements $Y_i^{(\ell)}$ as

$$T_M^2 = n \sum_{\ell=1}^{p} \sum_{\ell'=1}^{p} \left(\overline{Y}^{(\ell)} - \mu^{(\ell)}\right) S^{(\ell,\ell')}\left(\overline{Y}^{(\ell')} - \mu^{(\ell')}\right)$$

where $S^{(\ell,\ell')}$ is the (ℓ, ℓ')-th element of the matrix S^{-1}. A MINITAB program which computes the T^2-statistics has to invert the S-matrix computed above and to compute the quadratic form by multiplying the appropriate matrices. In Appendices 1.2-1.3 we present several macros for computing T^2-statistics for the various cases detailed in the next chapters.

Under the null hypothesis that the data are independent and normally distributed with the postulated mean vector μ, it follows immediately from (iv) above that the T_M^2-statistic just defined is distributed as

$$\frac{n - p}{(n - 1)p} T_M^2 \sim F_{p,n-p}.$$

When the covariance matrix was calculated from n observations and we want to assess the distance of a single observation $\mathbf{Y}$ from the expected value μ, the T_M^2 statistic is obtained as $T_M^2 = (Y - \mu)' S^{-1}(Y - \mu)$.

Another additional important special case is when $p = 1$. The T_M^2 statistic used for assessing the deviation of the mean of n observations from the expected value reduces in this case to the square of the t-statistic $t = \frac{\sqrt{n}(\overline{Y}-\mu)}{S}$ which is the statistic used for testing hypotheses on the mean of a univariate normal population. If the expected value of $\mathbf{Y}$ is μ, then the t-statistic has a Student t-distribution with $n-1$ degrees of freedom and since the coefficient of T_M^2 reduces to 1, then $t^2 \sim F_{1,n-1}$ as expected.

We illustrate the computations and the hypothesis with the first simulated data set presented above. We mentioned that the first 50 observations in the data set simulate data from an in-control process capability study. The underlying distribution was normal with

$$\mu = \mu_0 = \begin{bmatrix} 49.91 \\ 60.05 \end{bmatrix} .$$

We computed the T_M^2 values for those 50 observations and compared them with the appropriate critical values from the distribution of $\frac{(n-1)p}{n-p}$ $F_{p,n-p}$, which in this case (for $p = 2, n = 50$) are 13.97, 12.35 and 6.64 for $\alpha = .0027, \alpha = .005$ and $\alpha = .05$, respectively. The results are presented in the last column of Table 2.1. We mention that none of the T_M^2-values in the sample exceeded even the critical value at $\alpha = .05$. This result could have been expected, since we tested the distances of the empirical data from the location parameters of the distribution from which the data were generated. The only source of observed differences is thus random error. In the next sections, we shall return to this example, this time using an extended version of the data set, which simulated both a "base sample" as well as some "tested samples."

When a sample of kn independent normal observations are grouped into k rational subgroups of size n, then the distance between the mean $\overline{\mathbf{Y}}_j$ of the j-th subgroup and the expected values μ is computed by T_M^2 defined as

$$T_M^2 = n(\overline{\mathbf{Y}}_j - \mu)' S_p^{-1} (\overline{\mathbf{Y}}_j - \mu) .$$

Note that unlike the ungrouped case, the estimated covariance matrix is now pooled from all the k subgroups. Again, under the null hypotheses the data are independent and normally distributed with the postulated mean vector μ_1. It follows from (iv) and (viii) above that

$$\frac{k(n-1) - p + 1}{k(n-1)p} T_M^2 \sim F_{p,k(n-1)-p+1} .$$

In the case of the second set of simulated data with $p = 4$, $k = 50$, $n = 2$, the critical values for T_M^2 are 10.94, 18.17 and 20.17 for $\alpha = 0.5$, $\alpha = .005$ and $\alpha = .0027$, respectively. As we can see from Table 2.2, among the 50 subgroups, four of the T_M^2-values exceeded the critical

TABLE 2.1

Means with external values (49.91, 60.05)′—the parameters used to generate the data.
The S-matrix from the base sample (50 observations). The data are the base sample:

	VAR1	VAR2	T^2_M
1	49.8585	60.0008	1.6073
2	49.8768	59.9865	4.2488
3	49.8706	60.0055	1.0904
4	49.9117	60.0126	5.3767
5	49.8470	60.0165	4.5009
6	49.8883	60.0216	0.4419
7	49.9158	60.0517	0.2205
8	49.9152	60.0673	0.5629
9	49.9055	60.0726	2.2046
10	49.8969	60.0208	1.0951
11	49.9137	60.0928	4.9861
12	49.8586	59.9823	2.9972
13	49.9514	60.0866	1.9884
14	49.8988	60.0402	0.0122
15	49.8894	60.0720	5.5016
16	49.9403	60.0681	1.5848
17	49.9132	60.0350	1.2866
18	49.8546	60.0145	2.6554
19	49.8815	59.9982	2.6315
20	49.8311	59.9963	5.3803
21	49.8816	60.0457	1.8579
22	49.8501	59.9860	2.5277
23	49.9778	60.0875	7.0252
24	49.8690	60.0159	0.9596
25	49.8779	60.0055	1.2957
26	49.8680	60.0088	1.0088
27	49.9388	60.0711	1.2284
28	49.9133	60.0634	0.3785
29	49.9120	60.0560	0.1053
30	49.9250	60.0749	0.7337
31	49.9442	60.1100	3.8446
32	49.8386	59.9725	3.8154
33	49.9492	60.1014	2.4121
34	49.9204	60.0803	1.5392
35	49.8994	60.0625	1.5246
36	49.8703	60.0219	1.0618
37	49.8846	60.0271	0.2658
38	49.9580	60.0878	2.7294
39	49.8985	60.0329	0.1939
40	49.9397	60.0826	1.1778
41	49.8741	60.0061	1.0917
42	49.9140	60.0401	0.8206
43	49.9501	60.0810	2.0330
44	49.8865	60.0169	0.6997
45	49.8912	60.0406	0.2766
46	49.9252	60.0532	0.8988
47	49.9326	60.0741	0.7513
48	49.9680	60.1219	4.4314
49	49.9289	60.0709	0.5856
50	49.9233	60.0632	0.3528

TABLE 2.2

Means with external values (9.9863, 9.9787, 9.9743 and 14.9763)′—the parameters used to generate the data. The S_{pooled}-matrix from the base sample (50 groups of 2 observations). The data are the base sample:

Group	VAR1	VAR2	VAR3	VAR4	T^2_M
1	9.9765	9.9702	9.9673	14.9670	1.8495
2	9.9936	9.9851	9.9893	14.9813	8.5031
3	9.9870	9.9719	9.9693	14.9661	10.2566
4	9.9962	9.9866	9.9777	14.9856	3.5200
5	9.9774	9.9718	9.9651	14.9718	3.7810
6	9.9710	9.9683	9.9585	14.9638	7.5711
7	9.9813	9.9733	9.9729	14.9706	1.6984
8	9.9817	9.9774	9.9731	14.9723	1.1358
9	9.9887	9.9842	9.9810	14.9787	1.5410
10	9.9861	9.9789	9.9812	14.9761	3.7382
11	9.9997	9.9980	10.0001	14.9935	10.9735
12	10.0076	9.9969	9.9939	14.9985	7.7374
13	9.9725	9.9648	9.9553	14.9590	4.8619
14	9.9902	9.9814	9.9770	14.9781	0.6139
15	9.9786	9.9720	9.9669	14.9630	3.3072
16	9.9903	9.9893	9.9788	14.9900	12.4277
17	9.9975	9.9870	9.9872	14.9867	3.8160
18	9.9878	9.9881	9.9841	14.9813	5.3670
19	9.9879	9.9840	9.9774	14.9750	2.4178
20	9.9928	9.9839	9.9801	14.9826	0.7681
21	9.9819	9.9742	9.9688	14.9737	0.9384
22	9.9894	9.9813	9.9741	14.9772	0.7797
23	9.9891	9.9810	9.9751	14.9782	0.3178
24	9.9884	9.9783	9.9765	14.9722	4.0895
25	10.0112	10.0054	10.0015	15.0117	16.3026
26	9.9776	9.9686	9.9665	14.9683	1.8331
27	9.9893	9.9793	9.9793	14.9804	1.4995
28	9.9833	9.9761	9.9688	14.9705	0.8288
29	9.9975	9.9877	9.9816	14.9883	2.8201
30	9.9843	9.9760	9.9796	14.9826	6.2309
31	10.0107	10.0034	10.0013	15.0057	10.8461
32	9.9689	9.9619	9.9582	14.9452	19.6493
33	10.0061	9.9972	9.9930	14.9919	8.5064
34	9.9755	9.9673	9.9613	14.9686	4.2052
35	9.9928	9.9869	9.9832	14.9870	1.7318
36	9.9789	9.9735	9.9651	14.9739	5.2074
37	9.9645	9.9594	9.9503	14.9525	9.0034
38	10.0007	9.9934	9.9880	14.9876	4.7334
39	9.9844	9.9729	9.9694	14.9733	1.2992
40	9.9924	9.9830	9.9788	14.9822	0.7626
41	9.9909	9.9858	9.9800	14.9749	4.9135
42	9.9770	9.9684	9.9659	14.9612	4.4176
43	9.9923	9.9865	9.9851	14.9888	3.0087
44	9.9696	9.9651	9.9587	14.9650	7.9559
45	9.9861	9.9771	9.9773	14.9802	1.9193
46	9.9774	9.9714	9.9678	14.9749	5.5854
47	9.9757	9.9708	9.9690	14.9701	3.4430
48	9.9696	9.9621	9.9602	14.9630	5.9166
49	9.9809	9.9781	9.9687	14.9683	3.3282
50	9.9886	9.9809	9.9728	14.9796	1.4853

value at $\alpha = .05$ (for the 11th, 16th, 25th and 32nd subgroups). The T_M^2 for the 32nd subgroup exceeded the critical values at $\alpha = .005$ as well.

The methods of analysis described throughout this text tacitly assume that the distribution of the analyzed data is fairly well represented by the multivariate normal distribution. Procedures for testing the hypothesis of multivariate normality include both analytical methods and graphical techniques (e.g., Mardia, 1975; Koziol, 1993). When the data present clear evidence of lack of normality, we may try to transform the data using, for example, a multivariate generalization of the Box and Cox (1964) transformation. In the univariate case, the Box and Cox transformation improves the adherence to normality of a variable X by finding an appropriate constant λ and transforming X into $X^{(\lambda)}$ where

$$X^{(\lambda)} = \begin{cases} \frac{X^\lambda - 1}{\lambda} & \lambda \neq 0 \\ \log X & \lambda = 0 \text{ and } X > 0 \end{cases}$$

On the method for finding λ see Box and Cox (1964) or Box, Hunter and Hunter (1978). Andrews (1971) generalized this approach to the multivariate case by introducing a vector $\lambda = (\lambda_1, \ldots \lambda_p)$ and the corresponding transformed data vector

$$\mathbf{X}^{(\lambda)} = (X_1^{(\lambda_1)}, \ldots, X_p^{(\lambda_p)})'$$

which is constructed so that $\mathbf{X}^{(\lambda)}$ is approximately distributed as $N_p(\mu, \Sigma)$. In the univariate case we can find a transformation $g(X)$ that will transform data to approximate normally up to any level of precision. One should note, however, that in the multivariate case an appropriate transformation might not exist [Holland (1973)].

We thus continue to consider in the following chapters that either the original data are approximately normally distributed, or that the data have been preliminarily transformed and, on the new scale, the observations have, approximately, a multivariate normal distribution.

3

Quality Control with Externally Assigned Targets

Objectives:

The chapter provides methods of analysis for the case in which the targets are assigned externally and are not based on data analysis. Situations in which external targets are appropriate are discussed, and the computations involved are presented in detail using simulated data.

Key Concepts
- Target values
- Multivariate process control
- Multivariate hypothesis testing
- External values
- Tested sample
- Data grouping
- Decomposition of the Hotelling T^2-distance

We mentioned in Chapter 1 that the statistics involved in multivariate *process* control depend on the *"origins"* of the target values **m**. The methodology and the applications presented in this chapter are for the case when the target values are assigned externally, i.e. from an external requirement or when a prespecified standard $\mathbf{m}_0$ has to be met. It should be noted that comparisons with externally specified targets are not usually considered in statistical process control since such targets do not account properly for valid process characteristics. However, whenever multivariate prespecified standards are set, multivariate quality control procedures can be used to assess whether the multivariate means of the products equal the external targets $\mathbf{m}_0$. The next two chapters deal with cases in which target values are derived and computed internally. We shall distinguish between targets derived from the tested sample itself, and targets which are derived from a "reference" or "base" sample. If we denote by μ, the multivariate mean of the products **Y**, the statistical analysis in quality control with external targets is equivalent to testing hypotheses of the form

$$H_0 : \mu = \mathbf{m}_0$$

against

$$H_a : \mu \neq \mathbf{m}_0 .$$

Given a sample of size $\mathbf{n}_1$ from the population, we can compute $T_M^2 = n_1(\overline{\mathbf{Y}} - \mathbf{m}_0)'\mathbf{S}^{-1}(\overline{\mathbf{Y}} - \mathbf{m}_0)$. We have seen in the previous chapter that for normally distributed data, if $\mathbf{m}_0$ is the expected value of $\overline{\mathbf{Y}}$, then the statistic T_M^2 has a distribution given by $F_0 = \frac{n_1-p}{p(n_1-1)}T_M^2 \sim F_{p_1 n_1-p}$. Thus, under the null hypothesis mentioned above, $F_0 \sim F_{p_1,n_1-p}$. Under H_a, F_0 has a non-central F-distribution. The critical value for T_M^2 is thus

$$UCL = \frac{p(n_1 - 1)}{n_1 - p} \quad F_{p,n_1-p}^{\alpha}$$

where F_{p,n_1-p}^{α} is the upper 100% percentile of the central F-distribution with $(p, n_1 - p)$ degrees of freedom. If the value of T_M^2 exceeds the critical value, we conclude that the derivations between $\overline{\mathbf{Y}}$ and the external targets $\mathbf{m}_0$ cannot be explained by random error. In such cases an assignable cause affecting the process is to be suspected.

As a first example we shall use an extended version of the first simulated data set presented in the previous chapter. Now, in addition to the "base" sample of 50 observations presented in Chapter 2, we generated 25 bivariate observations which simulate a series of tested samples, whose parameters are as follows: the parameters of observations 51 – 55 are identical to those of the base sample (i.e. the data are in-control); the population mean for the first component in observations 56 – 65 has been shifted upward by two standard deviation i.e.

$$\mu_2 = \mu_0 + \begin{bmatrix} 2\sigma_1 \\ 0 \end{bmatrix} = \begin{bmatrix} 49.984 \\ 60.050 \end{bmatrix} ;$$

finally, in observations 66 – 75, the population mean of the first component was shifted downward and that of the second upward, both by one standard deviation, i.e.

$$\mu_3 = \mu_0 + \begin{bmatrix} -\sigma_1 \\ +\sigma_2 \end{bmatrix} = \begin{bmatrix} 49.873 \\ 60.087 \end{bmatrix} .$$

We recall that the first 50 bivariate observations were generated from the normal distribution with the mean vector

$$\mu_0 = \begin{bmatrix} 49.91 \\ 60.05 \end{bmatrix} .$$

We shall consider two sets of target values: the first set satisfies $m_0^{(1)} = \mu_0$, i.e. the means used for generating the data are taken as target values; in the second set, the target values are

$$m_0^{(2)} = \begin{bmatrix} 49.9 \\ 60.0 \end{bmatrix} .$$

We start by testing the hypothesis that the generated data originate from the population with the prespecified targets, i.e. $m = m_0^{(1)}$. If we further assume that the covariance matrix is known (as it is indeed the case in the simulated data), we can compute for each observation in the "tested" sample the statistic

$$\left(Y_i - m_0^{(1)} \right)' \Sigma^{-1} \left(Y_i - m_0^{(1)} \right) , \quad i = 51, \ldots, 75$$

and compare those values with the critical values from the chi-squared distribution with two degrees of freedom (see property (vi) in Chapter 2).

Alternatively, the covariance matrix Σ can be estimated from the "base" sample by the empirical covariance matrix S. The resulting T_M^2 statistics have then to be compared with the critical values based on the F-distribution which are presented in this chapter. We now perform for each of the 25 observations from the tested sample, (i.e, single observations $\mathbf{Y}$) both the multivariate T^2-test based on the T_M^2-statistic as well as the univariate t-test for each of the two components. We recall that we performed in Chapter 2 the corresponding multivariate tests for the first 50 observations from the "base" sample.

In the multivariate test, the null hypothesis is that the mean of the specific bivariate observation being tested satisfies:

$$\mathbf{m} = \mathbf{m}_0^{(1)} = \begin{bmatrix} 49.91 \\ 60.05 \end{bmatrix},$$

while in the univariate test, the hypotheses are stated for each component separately. The critical values for T_M^2 from the distribution of $\frac{2 \cdot 49}{48} F_{2,48}$ are 13.693, 12.103 and 6.515 for $\alpha = .0027$, $\alpha = .005$ and $\alpha = .05$, respectively. The corresponding critical values for the two sided t-test are $|t| = 3.16$, 2.94 and 2.01 for the values of α as above.

The results are presented in Table 3.1.

For each of the three subsets of the tested sample (of sizes 5, 10 and 10, respectively), we summarize in Table 3.2 the number of observations for which the null hypothesis has been rejected by the three statistics.

The first subset of five observations, which are in control, behave as expected and for none of them, the T_M^2 exceeded the appropriate critical values. In the second subset of 10 observations, only the mean of the first component has been shifted (by two standard deviations). We see that the empirical power of the multivariate T_M^2 test exceeded that of the univariate test performed specifically on the affected component. The single observation for which the second component exceeded the critical value at $\alpha = .05$ is to be considered a type I error since its mean has not been shifted.

Let us now consider the second set of target values (which are defined as the means used to generate the data but this time with only one significant digit), i.e.

TABLE 3.1

Means with external values (49.91, 60.05)' which are the parameters used
to generate the data. The S-matrix from the base sample (50 observations).
The data are the tested samples:

	VAR1	VAR2	t - VAR1	t - VAR2	T^2_M
51	49.8798	60.0417	-0.8654	-0.2258	1.6092
52	49.9208	60.0292	0.3100	-0.5635	3.6568
53	49.9606	60.1172	1.4490	1.8164	3.9386
54	49.9498	60.0543	1.1388	0.1153	5.6359
55	49.8390	59.9665	-2.0318	-2.2584	4.4074
56	49.9284	60.0079	0.5272	-1.1400	12.6205
57	49.9648	60.0482	1.5704	-0.0483	12.8573
58	49.9780	60.0186	1.9465	-0.8506	35.1905
59	50.0218	60.0854	3.2024	0.9562	27.9776
60	50.0606	60.1399	4.3127	2.4314	29.3890
61	50.0365	60.1005	3.6221	1.3651	30.3644
62	49.9756	60.0387	1.8781	-0.3067	22.3401
63	49.9840	60.0857	2.1186	0.9646	9.3746
64	50.0028	60.0482	2.6584	-0.0488	34.8288
65	49.9770	60.0278	1.9174	-0.6002	28.9178
66	49.8579	60.0588	-1.4911	0.2384	12.2043
67	49.8997	60.0820	-0.2938	0.8653	5.4627
68	49.9156	60.1415	0.1600	2.4750	24.2380
69	49.9258	60.1132	0.4538	1.7082	7.8701
70	49.8384	60.0449	-2.0487	-0.1381	15.6046
71	49.8937	60.0893	-0.4657	1.0619	9.5333
72	49.8631	60.0757	-1.3417	0.6960	16.8326
72	49.9406	60.1298	0.8750	2.1589	9.5522
74	49.9046	60.0739	-0.1554	0.6463	2.5946
75	49.8718	60.0676	-1.0931	0.4761	9.8310

$$\mathbf{m}_0^{(2)} = \begin{bmatrix} 49.9 \\ 60.0 \end{bmatrix}.$$

For the first 50 observations from the "base" sample as well as for the
first five observations in the "tested" sample, the targets deviate from
the presumably unknown means of the bivariate distribution by $0.27\sigma_1$
and $-1.35\sigma_2$, respectively. The deviations are more substantial for the
other "tested" observations.

The results of the testing for the 50 observations in the "base sample" with the covariance matrix being estimated by S, are presented

TABLE 3.2
Number of rejections of $H_0 : m = m_0^{(1)}$ in the three "tested" samples

	T_M^2			$|t|$-First Component			$|t|$-Second Component		
	$\alpha = .05$	$\alpha = .005$	$\alpha = .0027$	$\alpha = .05$	$\alpha = .005$	$\alpha = .0027$	$\alpha = .05$	$\alpha = .005$	$\alpha = .0027$
Observations 51-55	0	0	0	1	0	0	1	0	0
Observations 56-65	7	5	2	5	3	3	1	1	0
Observations 66-75	3	0	0	1	0	0	2	0	0

TABLE 3.3

Means with external values (49.9, 60.0)′. The S-matrix from the base sample (50 observations). The data are the base sample:

	VAR1	VAR2	t - VAR1	t - VAR2	T^2_M
1	49.8585	60.0008	-1.1881	0.0221	1.6073
2	49.8768	59.9865	-0.6646	-0.3639	4.2488
3	49.8706	60.0055	-0.8411	0.1482	1.0904
4	49.9117	60.0126	0.3349	0.3400	5.3767
5	49.8470	60.0165	-1.5169	0.4473	4.5009
6	49.8883	60.0216	-0.3339	0.5856	0.4419
7	49.9158	60.0517	0.4536	1.3974	0.2205
8	49.9152	60.0673	0.4354	1.8202	0.5629
9	49.9055	60.0726	0.1576	1.9625	2.2046
10	49.8969	60.0208	-0.0897	0.5621	1.0951
11	49.9137	60.0928	0.3913	2.5109	4.9861
12	49.8586	59.9823	-1.1861	-0.4778	2.9972
13	49.9514	60.0866	1.4729	2.3434	1.9884
14	49.8988	60.0402	-0.0342	1.0866	0.0122
15	49.8894	60.0720	-0.3022	1.9479	5.5016
16	49.9403	60.0681	1.1550	1.8434	1.5848
17	49.9132	60.0350	0.3788	0.9479	1.2866
18	49.8546	60.0145	-1.2992	0.3922	2.6554
19	49.8815	59.9982	-0.5303	-0.0485	2.6315
20	49.8311	59.9963	-1.9721	-0.1007	5.3803
21	49.8816	60.0457	-0.5279	1.2355	1.8579
22	49.8501	59.9860	-1.4297	-0.3777	2.5277
23	49.9778	60.0875	2.2265	2.3657	7.0252
24	49.8690	60.0159	-0.8881	0.4294	0.9596
25	49.8779	60.0055	-0.6324	0.1493	1.2957
26	49.8680	60.0088	-0.9156	0.2381	1.0088
27	49.9388	60.0711	1.1104	1.9224	1.2284
28	49.9133	60.0634	0.3821	1.7144	0.3785
29	49.9120	60.0560	0.3432	1.5150	0.1053
30	49.9250	60.0749	0.7151	2.0251	0.7337
31	49.9442	60.1100	1.2659	2.9759	3.8446
32	49.8386	59.9725	-1.7595	-0.7427	3.8154
33	49.9492	60.1014	1.4090	2.7436	2.4121
34	49.9204	60.0803	0.5831	2.1732	1.5392
35	49.8994	60.0625	-0.0177	1.6898	1.5246
36	49.8703	60.0219	-0.8496	0.5914	1.0618
37	49.8846	60.0271	-0.4414	0.7323	0.2658
38	49.9580	60.0878	1.6598	2.3742	2.7294
39	49.8985	60.0329	-0.0428	0.8896	0.1939
40	49.9397	60.0826	1.1365	2.2342	1.1778
41	49.8741	60.0061	-0.7428	0.1651	1.0917
42	49.9140	60.0401	0.4020	1.0851	0.8206
43	49.9501	60.0810	1.4337	2.1907	2.0330
44	49.8865	60.0169	-0.3857	0.4581	0.6997
45	49.8912	60.0406	-0.2515	1.0989	0.2766
46	49.9252	60.0532	0.7221	1.4392	0.8988
47	49.9326	60.0741	0.9335	2.0037	0.7513
48	49.9680	60.1219	1.9476	3.2971	4.4314
49	49.9289	60.0709	0.8285	1.9185	0.5856
50	49.9233	60.0632	0.6679	1.7102	0.3528

TABLE 3.4

Means with external values (49.9, 60.0)'. The S-matrix from the base sample (50 observations). The data are the tested samples:

	VAR1	VAR2	t - VAR1	t - VAR2	T^2_M
51	49.8798	60.0417	-0.5791	1.1267	1.6092
52	49.9208	60.0292	0.5962	0.7889	3.6568
53	49.9606	60.1172	1.7352	3.1689	3.9386
54	49.9498	60.0543	1.4250	1.4677	5.6359
55	49.8390	59.9665	-1.7455	-0.9059	4.4074
56	49.9284	60.0079	0.8135	0.2125	12.6205
57	49.9648	60.0482	1.8567	1.3041	12.8573
58	49.9780	60.0186	2.2327	0.5018	35.1905
59	50.0218	60.0854	3.4887	2.3086	27.9776
60	50.0606	60.1399	4.5989	3.7838	29.3890
61	50.0365	60.1005	3.9083	2.7175	30.3644
62	49.9756	60.0387	2.1644	1.0458	22.3401
63	49.9840	60.0857	2.4049	2.3170	9.3746
64	50.0028	60.0482	2.9447	1.3036	34.8288
65	49.9770	60.0278	2.2037	0.7522	28.9178
66	49.8579	60.0588	-1.2048	1.5908	12.2043
67	49.8997	60.0820	-0.0075	2.2177	5.4627
68	49.9156	60.1415	0.4463	3.8274	24.2380
69	49.9258	60.1132	0.7401	3.0606	7.8701
70	49.8384	60.0449	-1.7624	1.2144	15.6046
71	49.8937	60.0893	-0.1795	2.4143	9.5333
72	49.8631	60.0757	-1.0554	2.0484	16.8326
73	49.9406	60.1298	1.1613	3.5113	9.5522
74	49.9046	60.0739	0.1308	1.9988	2.5946
75	49.8718	60.0676	-0.8068	1.8285	9.8310

in Table 3.3. The corresponding results for the "tested sample" can be found in Table 3.4. The computations were performed by the macros presented in Appendix 1.2. Those macros compute the relevant statistics for the analyses of ungrouped data presented in Chapters 3–5. The program with the macros that compute the statistics for ungrouped data can be run by typing MTB>EXEC 'UU.MTB' 1.

The summary Table 3.5 presents the number of observations for which the null hypothesis has been rejected both for the "base" as well as for the various "tested" samples.

We observe again that the success of the univariate tests to detect deviations is much smaller than that of the multivariate test.

TABLE 3.5
Number of rejections of $H_0 : m = m_0^{(2)}$ in the "base" and in the three "tested" samples

| | T_M^2 | | | $|t|$-First Component | | | $|t|$-Second Component | | |
	$\alpha = .05$	$\alpha = .005$	$\alpha = .0027$	$\alpha = .05$	$\alpha = .005$	$\alpha = .0027$	$\alpha = .05$	$\alpha = .005$	$\alpha = .0027$
Observations 1-50	23	12	7	1	0	0	11	1	1
Observations 51-55	2	2	1	0	0	0	1	1	1
Observations 56-65	7	5	5	8	4	3	4	1	1
Observations 66-75	10	10	10	0	0	0	6	3	2

We mentioned in the introduction that the T_M^2 statistic does not provide the important information on the variable(s) which caused the out-of-control signal. On the other hand, we have seen in the example above that the power of the univariate t-test is lower than that of the multivariate test. Multiple univariate testings have a compounded problem when the significance level is assessed, due to the multiple comparisons issue. In Chapter 7 we present other methods for detecting the outlying variables. These methods are extensions of the multivariate testing and use the information from all the variables and their covariance structure.

As a second example of external targets we use the real data from Case Study 3, where the variables are dimensions of several lots of ceramic substrates. Raw materials used in the manufacturing of hybrid microcircuits consist of components, dyes, pastes, and ceramic substrates. The ceramic substrate plates undergo a process of printing and firing through which layers of conductors, dielectric, resistors, and platinum or gold are added to the plates. Subsequent production steps consist of laser trimming, component mounting and reflow soldering, or chip enwire bonding. The last manufacturing stage is the packaging and sealing of the completed modules. The ceramic substrates are produced in lots of varying sizes. The first production batch (labeled Reference) proved to be of extremely good quality yielding an overall smooth production with no scrap and repairs. This first lot was therefore considered a 'standard' to be met by all following lots. Five dimensions are considered in Case Study 3, with labels (a, b, c, W, L). The first three are determined by the laser inscribing process the last two are outer physical dimensions. In our next example we consider only the first set of three dimensions from the table from Appendix 2.

The Reference sample has average dimension, in mm, for $(a, b, c) = (199, 550.615, 550.923)$. The engineering nominal specifications are (200, 550, 550). By redefining the (a, b, c) measurements as the deviations from the nominal specification target values, a natural null hypothesis to test on the three dimensional population mean is:

$$H_0 : \mu = \begin{bmatrix} 0 \\ 0 \\ 0 \end{bmatrix}$$

TABLE 3.6
Measurements in reference sample (Case Study 3)

	a	b	c
Means ($\overline{Y}$)	-1.000	0.615	0.923
S^{-1}-matrix	1.081	0.100	-0.432
	0.100	1.336	0.063
	-0.432	0.063	2.620

against

$$H_a : \mu \neq \begin{bmatrix} 0 \\ 0 \\ 0 \end{bmatrix}$$

The means (with respect to the center of the nominal specifications) and the S^{-1} matrix of the Reference sample are as shown in Table 3.6.

Since the Reference sample has 13 units we have that $n = 13$ and

$$
\begin{aligned}
T_M^2 &= 13(\overline{Y} - 0)'S^{-1}(\overline{Y} - 0) \\
&= 13[-1.00\ 0.615\ 0.923] * \begin{bmatrix} 1.081 & 0.100 & -0.432 \\ 0.100 & 1.336 & 0.063 \\ -0.432 & 0.063 & 2.620 \end{bmatrix} * \begin{bmatrix} -1.000 \\ 0.615 \\ 0.923 \end{bmatrix} \\
&= 13[-1.427\ 0.793\ 2.890] * \begin{bmatrix} -1.003 \\ 0.615 \\ 0.923 \end{bmatrix} \\
&= 13 \times 4.58 \\
&= 59.54
\end{aligned}
$$

Now since $F_{3,12-3}^{.01} = 27.23$ we have to reject H_0 at a 1% level of significance. Therefore we conclude that, on the average, although the Reference sample was of extremely good quality it did not meet the required nominal specifications.

Kenett and Halevy (1984) investigated the multivariate aspects of military specifications documents published by the US Department of Defense. They show that a multivariate approach is indeed required for setting the inspection criteria of such products, since most product characteristics tend to be correlated. Again, such standards yield externally assigned targets that are used to monitor actual production.

Another source of externally assigned targets is the development process of a product or process. In such a development process one typically accounts for customer requirements, internal technical capabilities, and what the competition offers. The Quality Function Deployment (QFD) matrices mentioned in Chapter 4 below are used to integrate these different sources of information yielding agreed upon targets that are set so as to create "selling points" for the new product or process (Juran, 1988). Such targets are set on several dimensions and T_M^2 can be used to determine if these targets are met using the techniques presented here.

Grouping the Data.

If the tested sample of size n_1 is grouped in k rational subgroups of size n_j, $n_1 = \sum_{j=1}^{k} n_j$ the empirical covariance matrix is a pooled estimate from the k sample covariance matrices calculated from each subgroups, i.e. $S_p = \sum_{j=1}^{k}(n_j - 1)S_j \Big/ \sum_{j=1}^{k}(n_j - 1)$

Furthermore, if the subgroup sizes are all equal, i.e. $n_1 = kn$ then the pooled covariance matrix S_p is the average of the k individual matrices,

$$S_p = \sum_{j=1}^{k} S_j \Big/ k .$$

When we now test the mean of a subgroup j of n observations, the test statistic is defined as

$$T_{M_j}^2 = n\left(\overline{\mathbf{Y}}_j - \mathbf{m}_0\right)' S_p^{-1}\left(\overline{\mathbf{Y}}_j - \mathbf{m}_0\right)$$

where $\mathbf{Y}_j$ is the mean of the n observations in the subgroup. The critical value for testing the hypothesis that the j-th subgroup does not deviate from the targets $\mathbf{m}_0$ by more than random variation is given by:

$$UCL = \frac{pk(n - 1)}{k(n - 1) - p + 1} \; F_{p, k(n-1)-p+1}^{\alpha}$$

The grouping of the data also enables us to compute a measure of internal variability within the subgroup: For the j-th subgroup the measure of variability is given by

$$T_{Dj}^2 = \sum_{i=1}^{n}(\mathbf{Y}_{ij} - \overline{\mathbf{Y}}_j)'S_p^{-1}(\mathbf{Y}_{ij} - \overline{\mathbf{Y}}_j)$$

where $\mathbf{Y}_{ij}$ is the i-th observation in the j-th subgroup. When considering all the observations of the j-th subgroup, relative to the targets $\mathbf{m}_0$, one obtains a measure of overall variability T_{0j}^2 defined as

$$T_{0j}^2 = \sum_{i=1}^{n}\left(\mathbf{Y}_{ij} - \mathbf{m}_0\right)'S_p^{-1}\left(\mathbf{Y}_{ij} - \mathbf{m}_0\right).$$

From basic algebra we have that

$$T_{0j}^2 = T_{Mj}^2 + T_{Dj}^2.$$

The critical values for $T_{D_j}^2$ can be approximated by

$$UCL = (n-1)\chi_p^2(\alpha)$$

where $\chi_p^2(\alpha)$ is the $100 \cdot \alpha$-th percentile of the chi-squared distribution with p-degrees of freedom (see, e.g. Johnson,1985). The approximation is based on the substitution of S_p by Σ in the formula for the computation of $T_{D_j}^2$.

To illustrate the use of the test statistics we use an extended version of the second simulated data set with grouped data presented in Chapter 2. In addition to the first 100 observations (50 subgroups of size two), we generated 90 additional observations grouped in 45 subgroups of size two whose distributions were as follows: the parameters of observations 101-110 are identical to those of the base sample; the population mean for the first component of observations 110-130 have been shifted upward by two standard deviations, i.e.

$$\mu_2 = \mu_0 + \begin{bmatrix} 2\sigma_1 \\ 0 \\ 0 \\ 0 \end{bmatrix} = \begin{bmatrix} 10.0133 \\ 9.9787 \\ 9.9743 \\ 14.9763 \end{bmatrix},$$

in observations 131-150, the population mean of the first component was shifted downward and that of the second component upward, both by one standard deviation, i.e.

$$\mu_3 = \mu_0 + \begin{bmatrix} -\sigma_1 \\ +\sigma_2 \\ 0 \\ 0 \end{bmatrix} = \begin{bmatrix} 9.9728 \\ 9.9923 \\ 9.9743 \\ 14.9763 \end{bmatrix}.$$

In the fourth "tested" sample (observations 151-170) the shift was by one standard deviation in the first component with the sign of the deviation alternating in consecutive observations as follows: in the observations whose case number is odd, the mean of the first component was shifted downward, while in the observations whose case number is even, the shift was upward. Thus when the data are grouped in pairs, the last 10 groups formed from observations 151-170 have average mean whose expected values are as in the "base" sample, but the within group deviations are expected to be large. Finally, the population means for each of the four components in observations 171-190 were shifted upward by one standard deviation, i.e.

$$\mu_5 = \mu_0 + \begin{bmatrix} \sigma_1 \\ \sigma_2 \\ \sigma_2 \\ \sigma_3 \end{bmatrix} = \begin{bmatrix} 9.9998 \\ 9.9923 \\ 9.9888 \\ 14.9915 \end{bmatrix}.$$

The target values were set at

$$\mathbf{m}_0 = \begin{bmatrix} 9.98 \\ 9.98 \\ 9.98 \\ 14.98 \end{bmatrix}$$

which deviate from the population means by

$$\mathbf{m}_0 - \mu_0 = \begin{bmatrix} -0.47\sigma_1 \\ +0.10\sigma_2 \\ +0.39\sigma_3 \\ +0.24\sigma_4 \end{bmatrix}$$

We performed the tests for the 75 groups and compared the resulting T_M^2's and T_D^2's with the critical values at $\alpha = .05$, which are 10.94 and 9.49 respectively. The results are presented in Tables 3.7 and 3.8. The computations were performed by the macros presented in Appendix 1.3. Those macros compute the relevant statistics for the analyses of grouped data presented in Chapters 3–5. The program with the macros which compute the statistics for grouped data can be run by typing MTB>EXEC 'GG.MTB' 1.

We summarize in Table 3.9 the number of groups for which the appropriate T_M^2 and T_D^2-values exceeded the critical values in the "base" sample as well as in the four "tested" samples.

We can see from the table that even for the relatively small deviations between the targets and the actual means (by less than 0.5σ for each variable), the power of the multivariate test is about 60% for the groups in which the population means were not shifted. The detection is 100% in the second and the third "tested" samples and the power is again about 60%-70% in the last two "tested" samples. Note that in the last "tested" sample all the means were shifted by one standard deviation, but they were all shifted in the same direction. The probability of detection of such a shift is considerably smaller than in the case in which only two out of the four components are shifted by one standard deviation, *but* the shifts are in opposite directions. Obviously, a shift by one standard deviation of *all* four components in opposite directions (say in pairs) would have yielded even larger T_M^2-values.

The results in the column which summarizes the results for the T_D^2-tests illustrate the power of the test to detect within group variation in the fourth "tested" sample.

TABLE 3.7

Means with external values (9.98, 9.98, 9.98 and 14.98)′. The S_{pooled}-matrix from the base sample (50 groups of 2 observations). The data are the base sample:

Group	VAR1	VAR2	VAR3	VAR4	T^2_M	T^2_D	T^2_0
1	9.9765	9.9702	9.9673	14.9670	6.9126	11.5787	18.4917
2	9.9936	9.9851	9.9893	14.9813	18.4056	2.6110	21.0175
3	9.9870	9.9719	9.9693	14.9661	35.1165	2.3787	37.4961
4	9.9962	9.9866	9.9777	14.9856	18.8737	7.6355	26.5084
5	9.9774	9.9718	9.9651	14.9718	6.7657	3.1150	9.8803
6	9.9710	9.9683	9.9585	14.9638	9.1575	5.1778	14.3352
7	9.9813	9.9733	9.9729	14.9706	9.3300	1.8340	11.1650
8	9.9817	9.9774	9.9731	14.9723	5.9162	4.2786	10.1941
9	9.9887	9.9842	9.9810	14.9787	7.8017	0.2000	8.0003
10	9.9861	9.9789	9.9812	14.9761	9.8200	1.6535	11.4742
11	9.9997	9.9980	10.0001	14.9935	11.3981	4.6827	16.0805
12	10.0076	9.9969	9.9939	14.9985	22.6423	6.6440	29.2864
13	9.9725	9.9648	9.9553	14.9590	16.5223	0.9796	17.5028
14	9.9902	9.9814	9.9770	14.9781	13.5629	5.4059	18.9667
15	9.9786	9.9720	9.9669	14.9630	14.9679	2.7758	17.7426
16	9.9903	9.9893	9.9788	14.9900	10.5040	6.1168	16.6199
17	9.9975	9.9870	9.9872	14.9867	16.6040	2.4932	19.0972
18	9.9878	9.9881	9.9841	14.9813	4.3820	0.1646	4.5464
19	9.9879	9.9840	9.9774	14.9750	11.8774	5.1691	17.0457
20	9.9928	9.9839	9.9801	14.9826	12.4014	3.6190	16.0202
21	9.9819	9.9742	9.9688	14.9737	8.4692	2.2632	10.7336
22	9.9894	9.9813	9.9741	14.9772	14.1076	0.4522	14.5589
23	9.9891	9.9810	9.9751	14.9782	12.0774	7.6379	19.7158
24	9.9884	9.9783	9.9765	14.9722	20.1141	4.6382	24.7498
25	10.0112	10.0054	10.0015	15.0117	19.5601	3.5733	23.1334
26	9.9776	9.9686	9.9665	14.9683	9.1897	2.2299	11.4198
27	9.9893	9.9793	9.9793	14.9804	10.8875	1.8949	12.7836
28	9.9833	9.9761	9.9688	14.9705	12.7171	2.4218	15.1385
29	9.9975	9.9877	9.9816	14.9883	16.2834	3.5760	19.8582
30	9.9843	9.9760	9.9796	14.9826	5.5541	0.9283	6.4825
31	10.0107	10.0034	10.0013	15.0057	19.4523	1.0866	20.5401
32	9.9689	9.9619	9.9582	14.9452	36.0729	1.1480	37.2211
33	10.0061	9.9972	9.9930	14.9919	26.2016	3.3385	29.5403
34	9.9755	9.9673	9.9613	14.9686	9.8866	5.2834	15.1706
35	9.9928	9.9869	9.9832	14.9870	6.8393	4.0766	10.9169
36	9.9789	9.9735	9.9651	14.9739	8.4442	7.7542	16.1987
37	9.9645	9.9594	9.9503	14.9525	15.3261	3.8448	19.1714
38	10.0007	9.9934	9.9880	14.9876	19.4869	14.6015	34.0904
39	9.9844	9.9729	9.9694	14.9733	14.3155	0.8168	15.1326
40	9.9924	9.9830	9.9788	14.9822	12.9165	7.6439	20.5611
41	9.9909	9.9858	9.9800	14.9749	18.3518	3.0003	21.3509
42	9.9770	9.9684	9.9659	14.9612	16.8827	8.5347	25.4177
43	9.9923	9.9865	9.9851	14.9888	5.0976	7.0051	12.1022
44	9.9696	9.9651	9.9587	14.9650	7.3918	1.1271	8.5188
45	9.9861	9.9771	9.9773	14.9802	7.1717	2.1927	9.3647
46	9.9774	9.9714	9.9678	14.9749	5.0018	7.9634	12.9645
47	9.9757	9.9708	9.9690	14.9701	3.4050	4.1985	7.6042
48	9.9696	9.9621	9.9602	14.9630	8.0880	5.0988	13.1861
49	9.9809	9.9781	9.9687	14.9683	11.2887	0.7487	12.0375
50	9.9886	9.9809	9.9728	14.9796	11.9922	2.4071	14.3999

TABLE 3.8

Means with external values (9.98, 9.98, 9.98 and 14.98)'. The S_{pooled}-matrix from the base sample (50 groups of 2 observations). The data are the tested samples:

Group	VAR1	VAR2	VAR3	VAR4	T^2_M	T^2_D	T^2_0
51	9.9752	9.9631	9.9629	14.9669	13.4041	4.3124	17.7152
52	9.9948	9.9782	9.9728	14.9831	29.1965	1.1444	30.3390
53	9.9825	9.9780	9.9706	14.9844	10.2608	5.1232	15.3835
54	9.9808	9.9748	9.9664	14.9717	9.4387	9.0740	18.5131
55	9.9986	9.9819	9.9791	14.9747	49.8481	1.3336	51.1806
56	10.0136	9.9778	9.9720	14.9673	194.9893	2.4845	197.4676
57	10.0068	9.9729	9.9661	14.9723	134.1977	12.4554	146.6554
58	10.0189	9.9765	9.9786	14.9802	198.7654	15.6996	214.4651
59	10.0184	9.9872	9.9854	14.9924	110.2353	1.0674	111.3039
60	10.0124	9.9818	9.9746	14.9797	121.0134	2.5841	123.5975
61	10.0094	9.9789	9.9695	14.9769	118.8845	7.5874	126.4702
62	9.9974	9.9655	9.9548	14.9616	126.4785	5.2893	131.7633
63	10.0282	10.0023	9.9959	14.9947	130.0325	6.7462	136.7787
64	10.0127	9.9794	9.9784	14.9758	145.0159	2.4281	147.4440
65	10.0042	9.9657	9.9550	14.9619	180.6132	3.7681	184.3891
66	9.9817	9.9977	9.9808	14.9881	34.8933	5.4240	40.3167
67	9.9650	9.9865	9.9723	14.9711	43.3239	5.9636	49.2889
68	9.9639	9.9832	9.9709	14.9621	36.0596	9.1947	45.2506
69	9.9768	9.9922	9.9809	14.9819	23.3029	6.5072	29.8100
70	9.9711	9.9946	9.9759	14.9797	57.9523	7.0179	64.9679
71	9.9722	9.9902	9.9738	14.9736	34.7953	3.0990	37.8982
72	9.9784	9.9907	9.9761	14.9767	20.5829	3.6764	24.2601
73	9.9727	9.9880	9.9719	14.9796	33.2976	5.5188	38.8147
74	9.9583	9.9834	9.9624	14.9646	69.4106	3.3426	72.7517
75	9.9877	10.0065	9.9825	14.9964	63.4783	3.8257	67.3056
76	9.9769	9.9687	9.9704	14.9653	11.8044	36.0720	47.8761
77	9.9789	9.9682	9.9590	14.9656	19.2991	30.5845	49.8822
78	9.9901	9.9798	9.9719	14.9830	15.7072	43.6885	59.3950
79	9.9720	9.9648	9.9620	14.9607	10.3572	22.2830	32.6401
80	9.9736	9.9625	9.9596	14.9666	12.9881	14.8953	27.8834
81	9.9839	9.9759	9.9705	14.9793	8.4083	10.2363	18.6437
82	9.9819	9.9740	9.9705	14.9731	7.8902	18.0869	25.9764
83	9.9916	9.9845	9.9804	14.9792	12.9792	24.5291	37.5092
84	9.9851	9.9818	9.9718	14.9793	8.5208	28.8474	37.3676
85	9.9715	9.9642	9.9610	14.9553	17.5724	21.9927	39.5664
86	9.9962	9.9875	9.9802	14.9804	22.3014	3.9807	26.2821
87	10.0015	9.9960	9.9925	14.9925	13.7984	5.4804	19.2793
88	10.0056	10.0059	9.9992	15.0061	14.0653	5.7841	19.8486
89	10.0085	10.0020	10.0011	15.0012	18.2087	9.0957	27.3036
90	9.9945	9.9838	9.9811	14.9775	24.4965	1.2341	25.7311
91	9.9881	9.9830	9.9783	14.9786	7.5604	5.6353	13.1958
92	10.0039	9.9915	9.9921	14.9909	26.0641	2.8166	28.8802
93	10.0003	10.0019	9.9945	14.9986	10.5187	1.3998	11.9183
94	9.9878	9.9830	9.9778	14.9756	10.7594	0.1619	10.9214
95	9.9972	9.9861	9.9894	14.9893	15.1882	17.4943	32.6826

TABLE 3.9

Number of times that T_M^2 and T_D^2 exceeded the critical values

	T_M^2	T_D^2	Number of groups
Groups 1-50 – "Base" sample	29	2	50
Groups 51-55 – First "tested" sample	3	0	5
Groups 56-65 – Second "tested" sample	10	2	10
Groups 66-75 – Third "tested" sample	10	1	10
Groups 76-85 – Fourth "tested" sample	6	10	10
Groups 86-95 – Fifth "tested" sample	7	1	10

4

Quality Control with Internal Targets–Multivariate Process Capability Studies

Objectives:

The chapter provides examples of process capability studies carried out on multivariate data. The different steps in performing such studies are illustrated. Formulas for setting control limits as well as multivariate capability indices computed from the analyzed data are presented. The chapter also presents guidelines for interpreting T^2-charts and capability indices.

Key Concepts
- Reference sample
- Process capability study
- Statistical Control Charts
- Multivariate process capability indices
- Internal targets with the "Leave one out" approach
- Data grouping
- Base Sample

During an ongoing industrial process, quality control is typically performed with target values derived from a standard "reference" sample whose units have been determined to be of acceptable quality on all the analyzed variables. However, at the initial stage of a new or changed process, a thorough investigation of its characteristics and capabilities is required. In a process capability study, there typically are no preexisting data on the process characteristics. In particular, no target values based on prior information from the production of the component are available, and the target values therefore have to be calculated internally. It is important to distinguish between product specification limits which are derived from customer needs, and process quality characteristics which depend upon the processes involved in the production and delivery of the product. Univariate process characteristics are typically given in terms of an average, estimating the process mean, and a standard deviation, estimating the process variability. In the multivariate case the vector of means replaces the univariate process mean and the correlations between variables are added to the variables' standard deviations or variances.

The Western Electric *Statistical Quality Control Handbook* (1956) defines the term "process capability study" as "...the systematic study of a process by means of statistical control charts in order to discover whether it is behaving naturally or unnaturally; plus investigation of any unnatural behavior to determine its cause, plus action to eliminate any of the unnatural disturbance."

Process capability studies are therefore much more than a simple exercise of just estimating the means and the covariance matrix.

Some of the steps in a multivariate process capability study include:

1. Determination of the boundaries of the process selected to be studied and of the variables characterizing the process outputs. This step is typically the responsibility of management.

2. Determination of representative time frames for data collection to define the sampling process and the rational subgroups. The design of the data collection system has to be tailored to the variables characterizing the process outputs. This step requires in-depth knowledge of the process as it is actually being operated and of future needs and plans that might affect the process.

3. Performance of a cause and effect analysis linking process outputs characteristics to internal parameters and control factors. The tools used in this phase can include the simple and efficient fishbone diagram or the more comprehensive Quality Function Deployment (QFD) matrices. The analysis may be further refined and confirmed with statistically designed experiments. At this stage one can conduct experiments using concepts of robust design in order to make the best use of control factors as countermeasures to noise factors affecting the process. For more details on robust designs see, for example, Phadke (1989), and Kenett and Zacks (1997).

4. Data collection and analysis using univariate control chart on the individual variables, multivariate control charts on a combination of variables, and various statistical and graphical methods to investigate the structure displayed by the data.

5. Elimination of special causes of variability. This step requires evaluation of stability of the process and an identification of sources of variability such as measurement devices, work shifts, operators, batches or individual item.

6. Evaluation of the underlying probability model for the process, including checking for multivariate normality. When needed, transformations may be used to induce normality.

7. Computation of process performance indices or process capability indices. Process performance indices represent a measure of historical performance and are calculated using all of the data without considering stability over time. They are used to summarize past performance. Process capability indices are computed when data is collected over a short time frame (say 30 observations) and are used as predictors of process capability. Such indices require process stability, representative samples, normality of the process distribution and independence between the collected observations. Recently several authors suggested using multivariate process capability indices (e.g. Chan et al., 1991; Wierda, 1993).

The multivariate process capability study is a classical situation in which quality control is performed with internal targets. The vector of target values **m** is computed from the data (after the exclusion of outlying subgroups), and each observation or mean of subgroup in the process is compared with those target values.

When the data are ungrouped and the empirical covariance matrix S is based on the entire sample of n_1 observations, in the T_M^2 statistic for the i-th observation given by

$$T_M^2 = \left(\mathbf{Y}_i - \overline{\overline{\mathbf{Y}}}\right)' S^{-1} \left(\mathbf{Y}_i - \overline{\overline{\mathbf{Y}}}\right),$$

the statistics $\left(\mathbf{Y}_i - \overline{\overline{\mathbf{Y}}}\right)$ and S are not independently distributed. However it can be shown that $(n-1)S$ can be decomposed as $(n-1)S = (n-2) S_1 + (\mathbf{Y}_i - \overline{\overline{\mathbf{Y}}})(\mathbf{Y}_i - \overline{\overline{\mathbf{Y}}})'$ such that $(n-2)S_1$ has a Wishart distribution and S_1 is independent of $(\mathbf{Y}_i - \overline{\overline{\mathbf{Y}}})$. See e.g. Wierda (1994). Therefore, the distribution of T_M^2 does not follow anymore (up to a constant) a Fisher F distribution but rather a beta distribution (see property (v) in Chapter 2). The appropriate UCL is given by

$$UCL = (n_1 - 1) B_\alpha(p/2, (n_1 - p - 1)/2)$$

where $B_\alpha(\cdot, \cdot)$ is the upper $100 \cdot \alpha$-th percentile of the beta distribution with the appropriate number of degrees of freedom.

If we ignore the dependence between $\left(\mathbf{Y}_i - \overline{\overline{\mathbf{Y}}}\right)$ and S, we obtain (as suggested for example by Jackson (1959), Ryan (1988), Tracy et al. (1992)) an approximated UCL given by

$$UCL^* = \frac{p(n_1 - 1)}{n_1 - p} F_{p, n_1 - p}.$$

However, lately Wierda (1994) compared the values of UCL and UCL* for $\alpha = .005$; $p = 2, 5, 10$ and for various values of n_1. He found that the differences between UCL and UCL* are very substantial and the use of the approximated critical value is clearly unwarranted.

It is however possible to define for the capability study T^2-type statistics for which the statistics which measure the distance of the tested observation from the mean are independent of the estimated covariance matrix and whose critical values are based on the percentiles of the F-distribution. Two main alternatives were suggested in the literature: The first approach suggested by Wierda (1994) who considers

$$T_M^{2'} = \left(\mathbf{Y}_i - \overline{\overline{\mathbf{Y}}}\right)' S_{(-i)}^{-1} \left(\mathbf{Y}_i - \overline{\overline{\mathbf{Y}}}\right),$$

where only the covariance matrix is based on the $n_1 - 1$ observations but $\overline{\overline{\mathbf{Y}}}$ is based on the entire sample of size n_1. The critical value for this statistic is

$$UCL' = \frac{(n_1 - 1)(n_1 - 2)p}{n_1(n_1 - p - 1)} F_{p, n_1 - p - 1}^{\alpha}.$$

The second approach was suggested by Bruyns (1992) who considers the "Leave One Out" method and defines the statistic

$$T_M^{2''} = \left(\mathbf{Y}_i - \overline{\overline{\mathbf{Y}}}_{(-i)}\right)' S_{(-i)}^{-1} \left(\mathbf{Y}_i - \overline{\overline{\mathbf{Y}}}_{(-i)}\right)$$

where $\overline{\overline{\mathbf{Y}}}_{(-i)}$, $S_{(-i)}$ are the vector of the means and the covariance matrix, respectively, calculated from all but the i-th observation.

The critical value for that statistic is

$$UCL'' = \frac{n_1(n_1 - 2)p}{(n_1 - 1)(n_1 - p - 1)} F_{p, n_1 - p - 1}^{\alpha}.$$

It can be shown that for each observation i, $T_M^{2'}$ can be obtained as a function of T_M^2 as follows:

$$T_{M_i}^{2'} = \frac{(n_1 - 2)T_{M_i}^2}{(n_1 - 1) - [n_1/(n_1 - 1)]T_{M_i}^2}$$

This relationship can be very useful since it enables us to compute $T_M^{2'}$ without recomputing the covariance matrix for each observation separately. The critical values for $T_M^{2'}$ are based on the F-distributions which is more readily available than the beta distribution. Moreover, since the function which relates $T_{M_i}^2$ with $T_{M_i}^{2'}$ is monotonic, the critical values can be obtained directly by the relationship. Thus

$$UCL' = \frac{(n_1 - 2)UCL}{(n_1 - 1) - [n_1/(n_1 - 1)]UCL}$$

and

$$UCL = \frac{(n_1 - 1)UCL'}{(n_1 - 2) + [n_1/(n_1 - 1)]UCL'}$$

Wierda (1994) presents the three statistics and mentions that he finds the regular T_M^2 preferable, since on one hand it resembles more closely the T_M^2-statistics computed in other situations and on the other hand it avoids the need to compute several times either the covariance matrix (for $T_M^{2'}$) or both the covariance matrix and the mean vector (for $T_M^{2''}$). The disadvantage is obviously the need to appeal to the critical values from the beta distribution.

For methodological practical reasons we recommend a different approach than that suggested by Wierda. Since, we use T_M^2 with internal targets in the process capability study, in our opinion it is advisable to compare each observation with a statistic on which that observation had no effect, i.e. the "Leave One Out" approach. The relatively cumbersome computation of $\overline{\overline{Y}}_{(-i)}$ and $S_{(-i)}$ for each i, can easily be implemented in this age of fast computers. The use of the critical values based on the F-distribution can be considered in this case an "extra bonus" but certainly not a crucial consideration.

While as mentioned, the "Leave One Out" approach is preferable from the methodological point of view, in practical conditions the differences between the two methods is seldom noticeable. Unless the number of observations is very small, the effect of a single observation on the computed statistics is rarely such that the outlier is detected by the "Leave One Out" approach and not detected by T_M^2.

For an illustration of the methods, let us now return to the first simulated data set with ungrouped data presented in Chapter 2. We consider only the first 50 observations as simulating data derived from a process capability study with unknown parameters. We recall that the first 55 observations were generated from the distribution with the same (μ_0, Σ) while the means of the next 20 observations were shifted. We of course assume that at this testing stage, the underlying parameters are unknown to the investigator, and proceed to test each observation separately versus the empirical overall mean of the first 50 observations. The covariance matrix is also estimated from the base sample.

We present in Table 4.1 the values of the statistics T_M^2, $T_M^{2'}$ and $T_M^{2''}$, respectively. The statistics were computed by the program whose main macro is UU.MTB from Appendix 1.2. Note that we have in fact only two distinct methods since there is a one-to-one correspondence between T_M^2 and $T_M^{2'}$. The critical values for $\alpha = .05$ are 5.76, 6.40 and 6.66 and for $\alpha = .005$ they are 9.69, 11.90 and 12.39, respectively. The critical values for T_M^2 are based on the beta distribution, while those for $T_M^{2'}$ and $T_M^{2''}$ are based on percentiles of the F-distribution. We observe that only one out of the 50 T_M^2-statistics exceeded the critical values at $\alpha = .05$ (observation 23) and that none of them exceeded the critical values for smaller significance levels. The T^2-value for observation 23 exceeded the appropriate critical values at $\alpha = .05$ for all the methods. If this was indeed a process capability study, it is doubtful if this observation would have been labeled as outlier. It is more likely that the entire sample would have been accepted and used as a reference sample for future testings. The observation 23 would have been considered in this case a random error (one out of 50, at $\alpha = .05$). From the data generation process, we know that this was indeed a random error.

Grouping the Data.

We mentioned that in the process capability stage, the estimation of the internal variability within the subgroups may be particularly important. The data has of course to be divided at this stage into rational subgroups, much the same as it is likely to happen in future stages of the production process.

Thus, we test each of the k subgroup means $\overline{\mathbf{Y}}_j$, $j = 1, ..., k$ against the overall mean $\overline{\overline{\mathbf{Y}}}$.

The statistic

$$T_M^2 = n(\overline{\mathbf{Y}}_j - \overline{\overline{\mathbf{Y}}})S_p^{-1}(\overline{\mathbf{Y}}_j - \overline{\overline{\mathbf{Y}}})$$

is compared to the UCL of

$$\text{UCL} = \left(\frac{p(k-1)(n-1)}{k(n-1) - p + 1} \right) F_{p,kn-k-p+1}^{\alpha}$$

TABLE 4.1

Means from the base sample $(49.9026, 60.0441)'$. The S-matrix from the base sample (50 observations). The data are the base samples:

	VAR1	VAR2	T^2_M	T'^2_M	T''^2_M
1	49.8585	60.0008	1.6073	1.6559	1.6962
2	49.8768	59.9865	4.2488	4.9924	4.7536
3	49.8706	60.0055	1.0904	1.0968	1.1379
4	49.9117	60.0126	5.3767	6.7211	6.1752
5	49.8470	60.0165	4.5009	5.3606	5.0657
6	49.8883	60.0216	0.4419	0.4318	0.4549
7	49.9158	60.0517	0.2205	0.2134	0.2259
8	49.9152	60.0673	0.5629	0.5529	0.5813
9	49.9055	60.0726	2.2046	2.3363	2.3570
10	49.8969	60.0208	1.0951	1.1018	1.1426
11	49.9137	60.0928	4.9861	6.0980	5.6755
12	49.8586	59.9823	2.9972	3.3018	3.2602
13	49.9514	60.0866	1.9884	2.0856	2.1156
14	49.8988	60.0402	0.0122	0.0117	0.0124
15	49.8894	60.0720	5.5016	6.9263	6.3380
16	49.9403	60.0681	1.5848	1.6310	1.6716
17	49.9132	60.0350	1.2866	1.3059	1.3483
18	49.8546	60.0145	2.6554	2.8763	2.8674
19	49.8815	59.9982	2.6315	2.8470	2.8393
20	49.8311	59.9963	5.3803	6.7271	6.1804
21	49.8816	60.0457	1.8579	1.9366	1.9719
22	49.8501	59.9860	2.5277	2.7209	2.7213
23	49.9778	60.0875	7.0252	9.6859	8.3933
24	49.8690	60.0159	0.9596	0.9596	0.9988
25	49.8779	60.0055	1.2957	1.3157	1.3581
26	49.8680	60.0088	1.0088	1.0110	1.0510
27	49.9388	60.0711	1.2284	1.2436	1.2859
28	49.9133	60.0634	0.3785	0.3688	0.3894
29	49.9120	60.0560	0.1053	0.1014	0.1077
30	49.9250	60.0749	0.7337	0.7262	0.7600
31	49.9442	60.1100	3.8446	4.4224	4.2628
32	49.8386	59.9725	3.8154	4.3822	4.2274
33	49.9492	60.1014	2.4121	2.5819	2.5905
34	49.9204	60.0803	1.5392	1.5806	1.6220
35	49.8994	60.0625	1.5246	1.5647	1.6064
36	49.8703	60.0219	1.0618	1.0667	1.1078
37	49.8846	60.0271	0.2658	0.2577	0.2726
38	49.9580	60.0878	2.7294	2.9671	2.9514
39	49.8985	60.0329	0.1939	0.1874	0.1986
40	49.9397	60.0826	1.1778	1.1895	1.2315
41	49.8741	60.0061	1.0917	1.0983	1.1393
42	49.9140	60.0401	0.8206	0.8155	0.8514
43	49.9501	60.0810	2.0330	2.1368	2.1652
44	49.8865	60.0169	0.6997	0.6915	0.7241
45	49.8912	60.0406	0.2766	0.2683	0.2839
46	49.9252	60.0532	0.8988	0.8963	0.9341
47	49.9326	60.0741	0.7513	0.7443	0.7786
48	49.9680	60.1219	4.4314	5.2581	4.9798
49	49.9289	60.0709	0.5856	0.5758	0.6047
50	49.9233	60.0632	0.3528	0.3434	0.3626

in order to detect outlying subgroups (e.g., Ryan, 1988). This test procedure can also be used when a "base" sample is being calibrated. If we identify l subgroups with $T_M^2 > $ UCL and determine an assignable cause that justifies their removal from the data set we can recompute $\overline{\overline{Y}}$ and S_p using the remaining $k - l$ subgroups. The Upper Control Limits (UCL) for a future subgroup of n observations is then set to be:

$$\text{UCL} = \left(\frac{p\,(k - l + 1)(n - 1)}{(k - l)\,n - k + l - p + 1} \right) F^{\alpha}_{p,(k-l)n-k+l-p+1}$$

However, even after removing obvious outliers, the testing of each of the subgroup means against the overall mean should be performed with caution since possible remaining outlying observations or trends in the sample, may affect both the overall mean and the empirical covariance matrix. The subsequent tests for the individual subgroups could thus be biased.

It is noteworthy that the fact that the data are grouped does not necessarily imply that the estimated covariance matrix has to be pooled over all the groups. Specifically, both in this chapter as well as in the other chapters in this book, we focus on two main cases with respect to the grouping issue: either (a) the data are ungrouped, and then in the computation of T_M^2 we consider the distance of each observation from $\overline{\overline{Y}}$, i.e. $(Y_i - \overline{\overline{Y}})$, and use the empirical covariance matrix S, or (b) the data are grouped and for the j-th group we consider $(\overline{Y}_j - \overline{\overline{Y}})$ with the pooled covariance matrix S_p.

However, those are not the only possible methods of analysis. Indeed, even if the data are grouped in k groups of size n, if we believe that all the data obtained during the process capability study originate from the same distribution, the covariance matrix S (with $kn - 1$ degrees of freedom) is a more efficient estimator of Σ than Sp with $k(n-1)$ degrees of freedom (e.g. Alt (1982), Wierda (1994)). Some authors therefore suggest the use of the statistic

$$T_{Mj}^2 = n(\overline{Y}_j - \overline{\overline{Y}})' S^{-1}(\overline{Y}_j - \overline{\overline{Y}})'$$

The distribution of this statistic is similar to that presented for the ungrouped case, and we have

$$UCL = \frac{(kn-1)(k-1)}{n} B_\alpha(p/2, (kn-p-1)/2) .$$

In this case, we can also compute an alternative statistic which has, under the null hypothesis, an F-type distribution. The statistic is:

$$T^{2'}_{M_j} = n(\overline{\mathbf{Y}}_j - \overline{\overline{\mathbf{Y}}})' \widetilde{S}^{-1}_{(-j)} (\overline{\mathbf{Y}}_j - \overline{\overline{\mathbf{Y}}}) .$$

The definition of $\widetilde{S}_{(-j)}$ is more complicated than in the ungrouped case. There, $\widetilde{S}_{(-i)} = S_{(-i)}$, i.e. the covariance matrix could have been obtained directly from all but the i-th observation. In the grouped case we have

$$\widetilde{S}_{(-j)} = [(n-1)kS^2_p + n(k-1)S^*_{(-j)}]/(kn-2)$$

where $S^*_{(-j)}$ is the covariance matrix computed from the $k-1$ group means for all but the j-th group. The formula for $S^*_{(-j)}$ is thus

$$S^*_{(-j)} = \sum_{\ell \neq j} (\overline{\mathbf{Y}}_j - \overline{\overline{\mathbf{Y}}}_{(-j)})(\overline{\mathbf{Y}}_j - \overline{\overline{\mathbf{Y}}}_{(-j)})'$$

The critical value for $T^{2'}_M$ is

$$UCL' = \frac{(kn-2)(k-1)p}{k(kn-p-1)} F^\alpha_{p, kn-p-1} .$$

As in the ungrouped case there is no need to actually compute the $T^{2'}_{M_j}$ statistic directly. From the value of $T^{2'}_{M_j}$ we can compute

$$T^{2'}_{M_j} = \frac{(k-2)T^2_{M_j}}{(k-1)[k/(k-1)]T^2_{M_j}}$$

The use of the more efficient estimator of Σ (S instead of S_p) can be important during a process capability study when the number observations is limited. It is, however, obvious that the advantage gained by the use of a more efficient estimator of the covariance matrix may be counterbalanced by the bias induced if the assumption that the entire data set has the same distribution, is invalid. The values of the parameters are obviously unknown and in many cases slight shifts may occur in the process capability study which can effect the dispersion as well as the location. Therefore, we believe that in many cases it is unwarranted to assume the risk and we suggest that the use of S_p is more appropriate in practical situation as an estimator of the covariance matrix for grouped data, and we use it in our computations.

In the multivariate process capability studies, where we assess the characteristics of the process, the estimation of the internal variability within the subgroups may be especially important. The T_D^2 measure defined in Chapter 3 for the case when the targets are from a reference sample, can be similarly calculated for the j-th subgroup as

$$T_D^2 = \sum_{i=1}^{n}(\mathbf{Y}_{ij} - \overline{\mathbf{Y}}_j)'S_p^{-1}(\mathbf{Y}_{ij} - \overline{\mathbf{Y}}_j),$$

where $\mathbf{Y}_{ij}, i = 1, \ldots, n$ is the i-th observation in the j-th subgroup. The overall measure of variability $T_0^2 = T_M^2 + T_D^2$ provides the sum of the weighted distances of the n observations from the grand mean $\overline{\overline{\mathbf{Y}}}$.

We illustrate in Table 4.2 the results of the computations with the second simulated data set with four variables. The 100 observations from the simulated process capability study were grouped in 50 pairs. The grand mean $\overline{\overline{\mathbf{Y}}}$ and the pooled empirical covariance matrix are presented in Chapter 2. (Note that $\overline{\overline{\mathbf{Y}}}$ from this chapter equals $\overline{\overline{\mathbf{X}}}$ from Chapter 2). For each of the 50 subgroups, we compute T_M^2 and T_D^2 and compare them with the critical values. The statistics were computed by the program for grouped data whose main macro is GG.MTB from Appendix 1.3. The critical values for T_M^2 are 18.96, 17.08 and 10.28 for $\alpha = .0027, \alpha = .005$ and $\alpha = .05$, respectively. The statistic T_D^2 which assesses the within group variability is compared to the critical values approximated by $X_p^2(\alpha)$, that in this case are 16.25, 14.86 and 9.488 for the α-values as before. Among the 50 subgroups, we find five whose T_M^2-values exceeded 10.28, and one group whose T_M^2-value exceeded

TABLE 4.2

Means from the base sample (9.9864, 9.97932, 9.97525 and 14.9768)′.

The S_{pooled}- matrix from the base sample (50 Groups of 2 Observations).

The data are the tested samples:

Group	VAR1	VAR2	VAR3	VAR4	T^2_M	T^2_D	T^2_0
51	9.9752	9.9631	9.9629	14.9669	4.3124	5.6006	9.9132
52	9.9948	9.9782	9.9728	14.9831	1.1444	8.8938	10.0373
53	9.9825	9.9780	9.9706	14.9844	5.1232	13.2905	18.4131
54	9.9808	9.9748	9.9664	14.9717	9.0740	2.0769	11.1502
55	9.9986	9.9819	9.9791	14.9747	1.3336	19.4155	20.7486
56	10.0136	9.9778	9.9720	14.9673	2.4845	125.5682	128.0478
57	10.0068	9.9729	9.9661	14.9723	12.4554	77.3265	89.7838
58	10.0189	9.9765	9.9786	14.9802	15.6996	131.9750	147.6752
59	10.0184	9.9872	9.9854	14.9924	1.0674	62.2530	63.3211
60	10.0124	9.9818	9.9746	14.9797	2.5841	67.2113	69.7954
61	10.0094	9.9789	9.9695	14.9769	7.5874	65.1534	72.7396
62	9.9974	9.9655	9.9548	14.9616	5.2893	70.8581	76.1438
63	10.0282	10.0023	9.9959	14.9947	6.7462	76.6345	83.3807
64	10.0127	9.9794	9.9784	14.9758	2.4281	87.6476	90.0757
65	10.0042	9.9657	9.9550	14.9619	3.7681	112.6605	116.4350
66	9.9817	9.9977	9.9808	14.9881	5.4240	55.8683	61.2914
67	9.9650	9.9865	9.9723	14.9711	5.9636	72.9030	78.8694
68	9.9639	9.9832	9.9709	14.9621	9.1947	58.2018	67.3922
69	9.9768	9.9922	9.9809	14.9819	6.5072	45.6482	52.1554
70	9.9711	9.9946	9.9759	14.9797	7.0179	88.6448	95.6601
71	9.9722	9.9902	9.9738	14.9736	3.0990	55.3878	58.4912
72	9.9784	9.9907	9.9761	14.9767	3.6764	33.0686	36.7462
73	9.9727	9.9880	9.9719	14.9796	5.5188	56.2431	61.7600
74	9.9583	9.9834	9.9624	14.9646	3.3426	101.1653	104.5051
75	9.9877	10.0065	9.9825	14.9964	3.8257	85.0237	88.8508
76	9.9769	9.9687	9.9704	14.9653	36.0720	4.6509	40.7232
77	9.9789	9.9682	9.9590	14.9656	30.5845	4.5457	35.1300
78	9.9901	9.9798	9.9719	14.9830	43.6885	4.7208	48.4088
79	9.9720	9.9648	9.9620	14.9607	22.2830	3.7056	25.9887
80	9.9736	9.9625	9.9596	14.9666	14.8953	6.3224	21.2181
81	9.9839	9.9759	9.9705	14.9793	10.2363	3.5303	13.7670
82	9.9819	9.9740	9.9705	14.9731	18.0869	0.5604	18.6473
83	9.9916	9.9845	9.9804	14.9792	24.5291	1.2320	25.7615
84	9.9851	9.9818	9.9718	14.9793	28.8474	4.7872	33.6342
85	9.9715	9.9642	9.9610	14.9553	21.9927	6.7332	28.7265
86	9.9962	9.9875	9.9802	14.9804	3.9807	4.3015	8.2821
87	10.0015	9.9960	9.9925	14.9925	5.4804	4.4330	9.9134
88	10.0056	10.0059	9.9992	15.0061	5.7841	15.5645	21.3489
89	10.0085	10.0020	10.0011	15.0012	9.0957	9.0404	18.1360
90	9.9945	9.9838	9.9811	14.9775	1.2341	5.7927	7.0271
91	9.9881	9.9830	9.9783	14.9786	5.6353	0.3996	6.0351
92	10.0039	9.9915	9.9921	14.9909	2.8166	8.6158	11.4324
93	10.0003	10.0019	9.9945	14.9986	1.3998	12.2313	13.6305
94	9.9878	9.9830	9.9778	14.9756	0.1619	1.3657	1.5276
95	9.9972	9.9861	9.9894	14.9893	17.4943	5.3729	22.8668

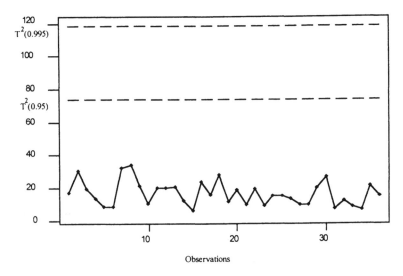

Figure 4.1: T^2-chart for the 17 variables from Case Study 2 (ungrouped).

18.98. In an actual capability study, such a subgroup should be excluded as a precaution. We know in this case that the data were simulated, and that the underlying distribution was identical to that of the other cases. However, in a real situation, such information is not available.

As for the within group variability in our data, we observe from the table that there were two T_D^2-values whose values exceeded 9.49 and one which exceeded 14.86. There was no T_D^2-value higher than 16.25. We mention that the seemingly outlying groups from the point of view of location (i.e. T_M^2) differ from those which have high internal variability. This result gives us a (weak) indication that there is a possibility that the outliers are in fact random errors.

Control Charts in Multivariate Process Capability Studies.
     After the deletion of the detected outliers, the process capability study just presented provides a "base" sample for future testing of the production process. We now illustrate the display for the T_M^2-statistic in a multivariate control chart for the ungrouped data from Case Study 2. The data have been collected in a process capability study of a complex mechanical part on which the 17 recorded variables include 11 diameters, 4 length and 2 wall-width measurements. The internal targets were

obtained from the 36 observations and the resulting T^2-chart is presented in Figure 4.1. The method of analysis is similar to that described in the section on analysis of control charts in the previous chapter. We only mention that there is no evidence of outlying observations in this particular data set.

While it is quite possible that in this particular case the process was in control throughout the process capability study a special feature of this data set deserves special attention. The large number of variables (17) relative to the number of observations (36) tends to favor the null hypothesis and to decrease the power of the test. It is thus recommended in similar situations to continue the data analysis on particular multivariate subsets of variables (possibly physically related) and investigate potentially assignable causes which may effect the studied process.

5

Quality Control with Targets from a Reference Sample

Objectives:

The chapter presents quality control procedures for comparing newly collected observations with targets determined from a reference sample. When the targets are based on a reference sample, the objective of multivariate quality control is to reach decisions on an on-going basis regarding the performance of a process over time as compared to the values from reference sample. This decision making process is illustrated with real life examples.

Key Concepts
- Reference sample
- Process capability study
- Tested sample
- Univariate control limits
- Multivariate control limits
- Data grouping
- Decomposition of the Hotelling T^2-distance

In the first phase of the process, i.e. the process capability study a "reference or base" sample is typically selected and based on its values, targets are set as reference for future testing of observations. Moreover, the data from the "reference" sample are used to assess the covariance pattern of the analyzed characteristics as well. In a second phase of the process the newly tested samples are routinely compared against the target values determined from the "reference" samples.

We note that comparison between a tested sample and the target values from a "reference" sample can also occur in the advanced stages of a process capability study. The selection of the "reference" sample out of the entire set of available data takes place as an initial stage of the process capability study. Following that stage, there may still be additional available observations which can be tested against the setup targets.

The statistical analysis for comparing a tested sample (whose observations are denoted by $\mathbf{Y}$) with a "reference" sample (denoted by $\mathbf{X}$) is based on the procedures for testing the equality of mean vectors of two parent populations. The sequential comparison of single observations or of subgroups of observations from the tested sample with the "reference" sample is an extension of the same procedures.

Let us assume that the tested sample which includes n_1 independent observations $\mathbf{Y}_1, \ldots, \mathbf{Y}_{n_1}, n_1 \geq 1$ distributed as $\mathbf{Y} \sim N_p(\mu_1, \Sigma_1)$. Furthermore, let us assume that here we also have a reference sample of n_2 independent observations $\mathbf{X}_1, \ldots, \mathbf{X}_{n_2}$ distributed as $\mathbf{X} \sim N_p(\mu_2, \Sigma_2)$. We assume throughout this chapter that $\Sigma_1 = \Sigma_2 = \Sigma$.

We start with the case where the samples ungrouped. Let $\overline{\mathbf{Y}}$ and $\overline{\mathbf{X}}$ be the vectors of averages for these two multivariate samples, respectively. We want to test the hypothesis $H_0 : \mu_1 = \mu_2$. From the results in Chapter 2, we have that $\overline{\mathbf{Y}} \sim N_p\left(\mu_1, \frac{1}{n_1}\Sigma\right)$ and $\overline{\mathbf{X}} \sim N_p\left(\mu_2, \frac{1}{n_2}\Sigma\right)$. And from the independence of $\overline{X}$ and $\overline{Y}$ we also have that

$$\overline{\mathbf{Y}} - \overline{\mathbf{X}} \sim N_p\left(\mu_1 - \mu_2, \left(\frac{1}{n_1} + \frac{1}{n_2}\right)\Sigma\right).$$

If Σ is known, we can calculate the statistic

$$T_M^2 = \frac{n_1 n_2}{n_1 + n_2} \left(\overline{\mathbf{Y}} - \overline{\mathbf{X}}\right)' \Sigma^{-1} \left(\overline{\mathbf{Y}} - \overline{\mathbf{X}}\right)$$

which follows a noncentral chi-square distribution with parameter of noncentrality

$$\tau^2 = \frac{n_1 n_2}{n_1 + n_2} (\mu_1 - \mu_2) \Sigma^{-1} (\mu_1 - \mu_2) .$$

When the null hypothesis $H_0 : \mu_1 = \mu_2$ holds, the parameter of non-centrality τ^2 is zero, and

$$T_M^2 \sim \chi_p^2 .$$

However, Σ is typically unknown and has to be estimated from the samples. Let S_X be the sample covariance matrix from the "reference" sample. If $n_1 \geq 2$, we can also calculate from the tested sample its covariance matrix S_Y. Based on S_X and S_Y, a pooled estimator of Σ can in principle be derived as

$$S = \left[(n_1 - 1)S_Y + (n_2 - 1)S_X\right] \Big/ (n_1 + n_2 - 2) .$$

If the quality control procedure calls for the comparison of consecutive single observations from the tested sample with the "reference" sample (i.e. $n_1 = 1$), Σ can obviously be estimated only from S_X. However, even when $n_1 \geq 2$, the information from the tested sample is rarely combined with that of the "reference" sample in the estimation of the covariance matrix. Since the comparison with the "reference" sample is performed routinely, it is much more convenient to keep the estimate of the covariance matrix constant when we compare the various tested samples with the "inference" samples. Furthermore, if the tested data contains outliers, they may differ from the parent population not only in their means, but also in their covariance structure.

Thus, the test statistic for $H_0 : \mu_1 = \mu_2$

$$T_M^2 = n_1 (\overline{\mathbf{Y}} - \overline{\mathbf{X}})' S_X^{-1} (\overline{\mathbf{Y}} - \overline{\mathbf{X}})$$

is compared to the Upper Control Limit (UCL) which is in this case

$$UCL = \frac{p(n_2 - 1)(n_1 + n_2)}{n_2(n_2 - p)} F_{p,n_2-p}$$

The control limits are used to determine departures from statistical control. A value of $T_M^2 > UCL$, or trends, or other patterns in the T_M^2-values, point out to a possible assignable cause affecting the process under observation. When a departure from statistical control is detected, one typically initiates an investigation to look for the specific characteristic that could be affected by that cause. Individual control charts are a logical next step in that direction. However, one should keep in mind that an aberrant T_M^2 value might be a result of observations whose correlation structures differ from that of the parent population and those features do not show up on the control chart for individual characteristics. Lately, Mason et al. (1995) have shown that the interpretation of a signal from a T_M^2-statistic is aided if the corresponding value is partitioned into independent parts. Information on which characteristic contributes significantly to the signal is readily available from that decomposition. We shall expand on that in the next chapter.

Let us now illustrate the methods of analysis with the simulated data set presented in Chapter 2 which includes both a "base" or "reference" sample of 50 observations as well as a "tested sample" with 25 observations. Unlike the analyses performed on that data set in the previous chapters, we now assume that both the vector of means as well as the covariance matrix are estimated from the "reference" sample. As mentioned in Chapter 2, we found in the "reference" sample

$$\bar{\bar{X}} = \begin{bmatrix} 49.9026 \\ 60.0441 \end{bmatrix}$$

and

$$S_X = \begin{bmatrix} .001220 & .001140 \\ .001440 & .001367 \end{bmatrix}$$

Using the macro UU.MTB from Appendix 3.2 with the properly defined parameters, we can compute the relevant statistics for each of the 25 single observations in the "tested sample" (i.e., $n_1 = 1$). We test the statistics obtained versus the UCL-values computed for various α's.

The results are presented in Table 5.1.

TABLE 5.1

Means from the base sample (49.9026, 60.0441)'. The S-matrix from the
base sample (50 observations). The data are the tested samples:

	VAR1	VAR2	t - VAR1	t - VAR2	T^2_M
51	49.8798	60.0417	-0.6532	-0.0656	1.6092
52	49.9208	60.0292	0.5222	-0.4033	3.6568
53	49.9606	60.1172	1.6612	1.9766	3.9386
54	49.9498	60.0543	1.3510	0.2754	5.6359
55	49.8390	59.9665	-1.8195	-2.0982	4.4074
56	49.9284	60.0079	0.7394	-0.9798	12.6205
57	49.9648	60.0482	1.7826	0.1119	12.8573
58	49.9780	60.0186	2.1587	-0.6905	35.1905
59	50.0218	60.0854	3.4146	1.1163	27.9776
60	50.0606	60.1399	4.5249	2.5915	29.3890
61	50.0365	60.1005	3.8343	1.5253	30.3644
62	49.9756	60.0387	2.0903	-0.1465	22.3401
63	49.9840	60.0857	2.3308	1.1247	9.3746
64	50.0028	60.0482	2.8706	0.1113	34.8288
65	49.9770	60.0278	2.1296	-0.4401	28.9178
66	49.8579	60.0588	-1.2789	0.3985	12.2043
67	49.8997	60.0820	-0.0816	1.0254	5.4627
68	49.9156	60.1415	0.3722	2.6351	24.2380
69	49.9258	60.1132	0.6660	1.8683	7.8701
70	49.8384	60.0449	-1.8365	0.0221	15.6046
71	49.8937	60.0893	-0.2535	1.2220	9.5333
72	49.8631	60.0757	-1.1295	0.8561	16.8326
73	49.9406	60.1298	1.0872	2.3190	9.5522
74	49.9046	60.0739	0.0568	0.8065	2.5946
75	49.8718	60.0676	-0.8809	0.6362	9.8310

We also summarize in Table 5.2 the number of observations for
which the statistics exceeded the appropriate critical values. The number of rejections are presented for each of the three subsets of the "tested
sample" (of sizes 5, 10 and 10, respectively). The results can be compared
with a similar table presented in Chapter 3. The difference between the
tables is that in Chapter 3 the null hypothesis specified the target value
of $\mathbf{m}_0$ while here we compare the $\mathbf{Y}'_i$'s with the empirical $\overline{\overline{\mathbf{X}}}$ obtained
from the "reference example". The critical values are changed accord-
ingly, and for the multivariate test they are 15.202, 12.346 and 6.645 for
$\alpha = .0027$, $\alpha = .005$ and $\alpha = .05$, respectively.

TABLE 5.2
Number of rejections in the three "tested" samples

	T_M^2			$|t|$-First Component			$|t|$=Second Component		
	$\alpha = .05$	$\alpha = .005$	$\alpha = .0027$	$\alpha = .05$	$\alpha = .005$	$\alpha = .0027$	$\alpha = .05$	$\alpha = .005$	$\alpha = .0027$
Observations 51-55	0	0	0	0	0	0	0	0	0
Observations 56-65	10	9	7	8	3	3	1	0	0
Observations 66-75	8	3	3	0	0	0	2	0	0

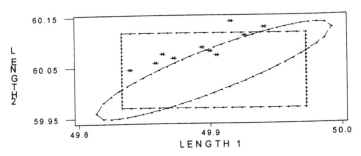

Figure 5.1: Scatterplot of ten bivariate observations from Case Study 1 with univariate and multivariate control limits.

The results in the table above are similar to those presented in Chapter 2 but the powers of the tests are higher in this case. We see again that the empirical power of T_M^2 exceeds that of the univariate test performed specifically on the effected component. An interesting pattern can be observed in the last subset of 10 observations from the tested sample (observations 66-75). The population means in those observations were all shifted by one standard deviations, but in opposite directions for the two components. In this case, the power of the multivariate test is substantial, while the univariate test detects almost no deviation from the standard. To further investigate the phenomenon, we present in Figure 5.1 the scatterplot of the data with the appropriate control limits. The interior and the boundaries of the ellipse is the area for which

$$T_M^2 \leq \frac{p(n_2 - 1)(n_2 + 1)}{n_2(n_2 - p)} F_{p,n_2-p}(.95) .$$

The rectangular lines represent the control limits for each of the two components separately. The control limits are set for the two sided test at $\alpha = .05$. The difference in the percent of rejections between the multivariate and univariate procedures is due to the fact that the multivariate test accounts for the dependence between the variable while the univariate test does not. Let us consider for example observations 70 and 72. The first component in case 70 is below its population mean by $1.845s_1$ and the second component by $0.025s_2$. In observation 72 the first component is below the mean by $1.13s_1$, and the second component is above its mean by $0.86s_2$. None of these marginal deviations exceeds the univariate critical limits, but given the correlation of $r = .883$, the probability of the events represented by the bivariate values would be very unlikely if the process were in control.

Grouping the Data.

Let us now turn to the case in which the samples are grouped in rational subgroups of size n. We denote in this case by $\overline{\overline{X}}$ the grand mean of the "reference" samples, and by S_{X_p} the pooled estimate covariance matrix based on k subgroups of size n.

When we now test the mean of a new group of n observations, (i.e. $n_1 = n$), the test statistic

$$T_M^2 = n(\overline{Y} - \overline{\overline{X}})' S_{X_p}^{-1} (\overline{Y} - \overline{\overline{X}})$$

is compared to the Upper Control Limit (UCL), which in this case is

$$UCL = \frac{p(k+1)(n-1)}{k(n-1) - p + 1} F_{p, kn-k-p+1}^{(\alpha)}$$

For a new subgroup of n observations $Y_i, i = 1 \ldots n$, with mean $\overline{Y}$ we can also compute the internal measure of variability T_D^2, i.e.

$$T_D^2 = \sum_{i=1}^{n} (Y_i - \overline{Y})' S_{X_p}^{-1} (Y_i - \overline{Y}) .$$

When we consider the deviations of all the observations in the new subgroup from the mean $\overline{\overline{X}}$ of the "reference" sample, we obtain the measure of overall variability T_0^2,

$$T_0^2 = \sum_{i=1}^{n} (Y_i - \overline{\overline{X}})' S_{X_p}^{-1} (Y_i - \overline{\overline{X}}) ,$$

with

$$T_0^2 = T_M^2 + T_D^2 .$$

We illustrate the computation of statistics presented for grouped data with a further analysis of the data from the capability study of aluminum pins (Case Study 1). We mentioned in the previous chapter that the 70 observations in that study were composed of two subsets: the first 30 observations were considered a "reference" sample against which we compare the other 40 observations. Furthermore, the data in both the "reference" and the tested sample, are grouped in subgroups of two consecutive observations (15 and 20 subgroups, respectively). Table 5.3 presents both the relevant statistics from the "reference" sample as well as the two observations (Y_1 and Y_2) of subgroup 16, which is the first subgroup in the tested sample. Those are the observations Y_{31} and Y_{32} in the original data set. The vector of means ($\overline{Y}$) is also presented. The statistics from the "reference" sample are the vector of means $\overline{\overline{X}}$ and the covariance matrix S_{X_p} (pooled over the 15 subgroups) and its inverse $S_{X_p}^{-1}$.

The T_M^2 – statistics from subgroup 16 is given by

$$T_M^2 = n(\overline{Y} - \overline{\overline{X}})' S_{X_p}^{-1} (\overline{Y} - \overline{\overline{X}})$$

which in our case is:

$$T_M^2 = 2*[0.0137\ 0.0113\ 0.0157\ 0.0187\ 0.0177\ -0.0177]$$

$$\begin{bmatrix} 75824.7 & -14447.7 & -29073.3 & -29977.6 & -4940.9 & 5301.9 \\ -14447.7 & 63781.5 & -31818.8 & -4551.4 & 709.5 & 856.4 \\ -29073.3 & -31818.8 & 61607.7 & -253.9 & 3508.6 & -3419.4 \\ -29977.6 & -4551.4 & -253.9 & 32532.6 & 1937.6 & -2630.0 \\ -4940.9 & 709.5 & 3508.6 & 1937.6 & 3172.4 & -3029.8 \\ 5301.9 & 856.4 & -3419.4 & -2630.0 & -3029.8 & 4158.8 \end{bmatrix} * \begin{bmatrix} 0.0137 \\ 0.0113 \\ 0.0157 \\ 0.0187 \\ 0.0177 \\ -0.0177 \end{bmatrix}$$

$$= 2*[-323.488\ -60.6431\ 324.8947\ 222.7156\ 141.2242\ -147.497] * \begin{bmatrix} 0.0137 \\ 0.0113 \\ 0.0157 \\ 0.0187 \\ 0.0177 \\ -0.0177 \end{bmatrix}$$

$$= 2 * 9.24$$

$$= 18.48$$

Note that unlike the calculations in the previous chapters, the targets are now the means $\overline{\overline{X}}$ of the "reference" sample. In Chapter 3, the targets were the external values m_0 while in Chapter 4 they were, at first, the

TABLE 5.3

Computation of the T_M^2-statistics – case study 1

	Diameter 1	Diameter2	Diameter3	Diameter4	Length1	Length2
$\mathbf{Y}_1$	10	9.99	9.99	14.99	49.92	60.03
$\mathbf{Y}_2$	10	9.99	9.99	15	49.93	60.03
$\overline{\overline{\mathbf{Y}}}$	10	9.99	9.99	14.995	49.925	60.03
$\mathbf{X}$	9.986	9.979	9.974	14.976	49.907	60.048
S_p-matrix(**10,000)						
Diameter1	1.10					
Diameter2	0.83	0.87				
Diameter3	0.97	0.87	1.10			
Diameter4	1.13	0.90	1.03	1.50		
Length1	-0.97	-1.20	-1.30	-1.10	12.33	
Length2	-0.77	-0.83	-0.80	-0.63	8.70	8.83
S_p^{-1}-matrix						
Diameter1	75824.7					
Diameter2	-14447.7	63781.5				
Diameter3	-29073.3	-31818.8	61607.7			
Diameter4	-29977.6	-4551.4	-253.9	32532.6		
Length1	-4940.9	709.5	3508.6	1937.6	3172.4	
Length2	5301.9	856.4	-3419.4	-2630.0	-3029.8	4158.8

internal values $\overline{\overline{Y}}$ based on the entire sample of 70 observations. Also, the $S_{X_p}^{-1}$ matrix presented above differs from the S_p^{-1} matrix used in the previous chapters since its calculation is based on 15 (rather than 35) subgroups. The T_M^2 for the same subgroup in Chapter 3 was 68.43.

The internal measure of variability $T_D^2 = \sum_{i=1}^{2} (Y_i - \overline{\overline{Y}})' S_{X_p}^{-1} (Y_i - \overline{\overline{Y}})$ is given in this case by

$$T_D^2 = [0\ 0\ 0\ -0.005\ -0.005\ 0] S_{X_p}^{-1}$$
$$*[0\ 0\ 0\ -0.005\ -0.005\ 0]'$$

$$+[0\ 0\ 0\ 0.005\ 0.005\ 0]S_{X_p}^{-1}$$
$$*[0\ 0\ 0\ 0.005\ 0.005\ 0]'$$
$$= 2*0.99$$
$$= 1.98$$

and the value of $T_0^2 = \sum_{i=1}^{2}(\mathbf{X}_i - \overline{\mathbf{Y}})'S_{X_p}^{-1}(\mathbf{X}_i - \overline{\mathbf{Y}})$ is

$$
\begin{aligned}
T_0^2 &= [0.0137\ 0.0113\ 0.0157\ 0.0137\ 0.0127\ -0.0177]S_{X_p}^{-1} \\
&\quad *[0.0137\ 0.0113\ 0.0157\ 0.0137\ 0.0127\ -0.0177]' \\
&\quad +[0.0137\ 0.0113\ 0.0157\ 0.0237\ 0.0227\ -0.0177]S_{X_p}^{-1} \\
&\quad *[0.0137\ 0.0113\ 0.0157\ 0.0237\ 0.0227\ -0.0177]' \\
&= 6.59 + 13.87 \\
&= 20.46
\end{aligned}
$$

It can easily be checked that $T_0^2 = T_M^2 + T_D^2$. The decomposition of the T_0^2-value into the components which measure the distance from target (T_M^2) and the internal variability (T_D^2), respectively, clearly indicates that the distance from target accounts for the main part of the overall variability (18.48 out of 20.46). The two observations which form subgroup 16 (observations 31 and 32), differ only slightly from each other, by .01 in the fourth and the fifth variables (see Table A2.1 in Appendix 2). The small T_D^2 value reflects this homogeneity within the 16th subgroup. We repeat the conclusion presented in Chapter 3 on the relative magnitude of T_M^2 and T_D^2 in this subgroup. The relatively large value of T_M^2 indicates a substantial deviation from the mean of the "reference" sample, while the small T_D^2-value indicate that the two observations in subgroup 16 are internally homogeneous.

6

Analyzing Data with Multivariate Control Charts

Objectives:

The chapter shows how to practically analyze data with multivariate control charts. The multivariate control chart is one of the main practical tools in multivariate quality control. The chart displays the T^2-distance with the upper control limits. The control limits are set at the critical values determined by the nature of the targets, as presented in the previous three chapters.

Key Concepts
- Graphical display of data
- Univariate control limits
- Critical values
- Multivariate control limits
- Data grouping
- The T^2-chart

The graphical display of data plays an important role in the detection of changes in the level of a variable characterizing a process. In fact, in many practical instances with a single attribute monitored, statistical quality control consists mainly of plotting the data on a control chart assessing levels and/or trends in the variable based on the graphical display, and *taking action* when it is determined that an assignable cause is present. Furthermore, even in the cases when several attributes are monitored, the most common practice is to monitor them on several separate displays. Thus, although we focus on multivariate observations, it is important to review briefly the univariate control charts which in many practical cases will at least supplement the multivariate charts.

The univariate control charts introduced by Shewhart in 1924 and described in his classic book of 1931 are considered a cornerstone in the development of statistical quality control in industry. Trends and changes in the levels of a variable are naturally and quickly perceived from a graphical display rather than from a listing of tabulated numbers. This general observation is particularly valid within an industrial setting.

The basic univariate Shewhart control charts successively display values of individual observations, or averages of subgroups of observations, and are termed respectively x-charts and $\bar{x}$-charts. When the data are grouped, the $\bar{x}$ charts are frequently accompanied by charts which display the value of a statistic which measures the dispersion within the groups (such as ranges or standard deviations). Examples of those are the R-charts and the SD-charts. The paired $(\bar{x}, R)$ charts are usually displayed and maintained on the production floor in industry to help monitor key process characteristics. We illustrate in Figure 6.1 the $(\bar{x}, R)$ chart with the data set on the sixth variable (overall length with cap or **Length2**) from Case Study 1 (35 subgroups of size 2).

From the chart we can observe that the overall length is significantly below target in subgroups 24 and 26 and so are, to a lesser extent, the two adjacent neighboring subgroups (23 and 27). The R-chart indicates that there are large dispersions between the two observations which form subgroups 25 and 31.

Various additional charts, for monitoring both process location and process dispersion, have been developed and used in industry. These include median charts, midrange, CUSUM, moving average, exponentially weighted moving average, etc. A detailed description of those univariate

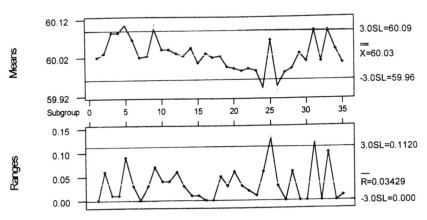

Figure 6.1: Shewart chart of $\overline{x}$-bar and range of Length 2 from Case Study 1 (2 observations per group).

control charts can be found in Wadsworth et al. (1986) and Kenett and Zacks (1997).

However, when several variables have to be monitored simultaneously, even if the univariate charts pertaining to all the relevant variables are created, these separate displays provide only partial information about the process. This is especially true when the quality characteristics are mildly or highly correlated. The pairwise scatterplots of the four diameter and two length measurements in Figure 1.1 indicate that this is indeed the case for the aluminum pins data. There are also obvious difficulties in simultaneously analyzing multiple charts. In particular, when tracking several dimensions simultaneously, the decision that an observation is out-of-control is subject to a higher rate of false alarms than for a single chart due to the multiple comparisons effect.

One approach to overcome these difficulties is to extend the univariate analysis by plotting a statistic which measures the overall deviations of the (multivariate) observations from the target. The most commonly used among such statistic is the Hotelling T_M^2 presented in the previous sections. The multivariate charts which display the Hotelling T_M^2-statistic are sometimes called Shewhart charts as well (e.g. Crosier, 1988), although Shewhart has not been involved in their development. As in the univariate case, when the data are grouped, the chart which displays the distance of the group means from the targets can be accompanied by a chart depicting a measure of dispersion within the subgroups for

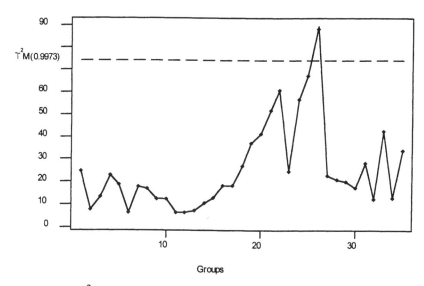

Figure 6.2a: T_M^2-chart for the 6 variables from Case Study 1 (2 observations per group).

all the analyzed characteristics such as the T_D^2-statistic presented above. The combined T_M^2 and T_D^2 charts thus provide a multivariate counterparts of the univariate $(\overline{x}, R)$ or $(\overline{x}, SD)$ charts and display for the multivariate data the *overall* measures of distance from the target and dispersion within the group. We present in Figure 6.2a the T_M^2 and in Figure 6.2b the T_D^2 charts for the data on all six variables from the capability study of aluminum pins. The vertical position of each plotted symbol corresponds to the value of T_M^2 (or T_D^2) for the respective subgroup of observations.

We observe from the multivariate control chart that starting at the 18-th subgroup and up to the 26-th subgroup the process shifts out of control (with the exception of the 23-rd subgroup). At the 26-th group the deviations from the target are large enough (at least on some variables) such that the overall T_M^2 value exceeds the critical 0.997 value. The process then seems to return to be in control (with the possible exception of group 33). We are not aware of any corrective action taken after the 26-th group, although we do not rule it out. The T_D^2-chart shows evidence of large dispersions between the two observations within subgroups 31 and 33. As in the analysis of the T_M^2-chart, we cannot assess, from the overall T_D^2-values, which dimensions display large variations.

The examination of the two T^2-charts is simpler than the examination of the six individual univariate Shewhart charts for each variable. Nevertheless, the multivariate Shewhart charts are considerably less popular than their univariate counterparts. Besides the relative difficulty involved in their computation, there are some additional features which diminish their appeal. First, unlike the univariate case, the scale of the values displayed on the chart is not related to the scales of any of the monitored variables. Secondly, when the T^2-statistic exceeds the critical value marked on the graph, the signal conveyed to the user lacks the relevant information on what might have caused it. Further investigation is required to determine which particular variables are responsible for the overall outlying value. This information is clearly essential in any attempt to detect the causes of the deviations and possibly correct the process. We present in Chapter 7 some recently developed analytical methods designed to address this problem and in Chapter 11 we present some graphical methods designed to detect the causes of deviations. The usefulness of the graphical methods presented in Chapter 11 is illustrated by reinvestigating the data which form the above mentioned outlying groups.

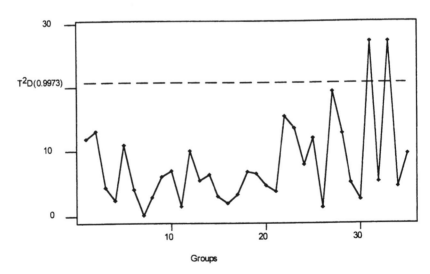

Figure 6.2b: T_D^2-chart for the 6 variables from Case Study 1 (2 observations per group).

7

Detection of Out-of-Control Characteristics

Objectives:

The chapter provides several multivariate methods for detecting the variables that cause the out-of-control signal. Special emphasis is devoted to the step-down procedure which is presented in detail.

Key Concepts
- Subsets of variables
- Step-down procedures
- The T^2 decomposition
- Regression adjusted variables

In the previous chapters, we presented the T_M^2-statistics and control charts for testing the hypotheses that a new multivariate observation (or group of observations) originates from the same distribution as that which generated the "target values." When the T_M^2-value signals that an observation is out of control, an important practical issue is to detect the variables which caused that signal. While it is common in industry to monitor individual process characteristics, the use of such procedures alone, is far from being satisfactory, since those procedures do not account for the correlations among the variables.

In recent years a series of multivariate control procedures was developed for detecting the variables which are out of control. In general the methods use various functions of decompositions of T_M^2-statistics. The T_M^2-statistics are either the overall statistics obtained from the initial analysis, or those obtained from the analysis of a subset of the p characteristics. Several methods are presented in this chapter. For the sake of simplicity, we restrict ourselves to the case when the target values are assessed from a "reference sample" of size n_2 and the "tested sample" is a new p-dimensional observation (i.e. $n_1 = 1$), but the extension to grouped data is straightforward.

As mentioned in Chapter 5, in this case the overall T_M^2-statistic is:

$$T_M^2 = (\mathbf{Y} - \overline{\mathbf{X}})' S^{-1} (\mathbf{Y} - \overline{\mathbf{X}})$$

and under the null hypothesis,

$$\frac{n_2(n_2 - p)}{p(n_2 - 1)} T_M^2 \sim F_{p, n_2 - p} \, .$$

When the T_M^2-value exceeds the critical value, we attempt to detect the variables responsible for that signal. The appropriate methods of analysis can be broadly divided into two groups:

(1) procedures which assume that there is an a priori ordering among subsets of the variables. The tests are performed successively. At each stage (except the first), we test conditional hypotheses for a specific subset. The conditioning is on that the means of the variables in the previous subsets are as in the reference population (step-down procedures),

(2) procedures which assume no specific ordering among the variables.

The step-down procedures.

The new observation $\mathbf{Y}$, whose T^2-value tested significant, has a distribution with mean μ_Y. Its covariance matrix Σ is assumed to be as in the reference population. We mentioned that the step-down procedure assumes an a priori ordering among the subsets of the variables. This ordering also governs the stages of the testings as well as the partitioning of the μ_Y-vector and of the Σ-matrix. Thus μ_Y is partitioned into q subsets denoted as $\mu_{y_1}, \ldots, \mu_{y_q}$, respectively, i.e.

$$
\mu_Y = \begin{bmatrix} \mu_{y_1} \\ \mu_{y_2} \\ \vdots \\ \mu_{y_q} \end{bmatrix}.
$$

The mean vector μ_X of the reference population is partitioned similarly into $\mu_{x_1}, \ldots, \mu_{x_q}$. Let us denote further subsets of μ_Y and μ_X as follows: μ_j^* includes all the variables belonging to all the original subsets from 1 to j (including). Similarly for $\mu_{x_j}^*$. Note that unlike the μ_{y_j}, μ_{x_j} subsets $(j = 1, \ldots, q)$, the $\mu_{y_j}^*$'s (and the $\mu_{x_j}^*$'s) are not mutually exclusive. Let p_j be the number of variables in the j-th subset, satisfying $\sum_{j=1}^{q} p_j = p$, and let $q_j = \sum_{k=1}^{j} p_j$ be the number of variables in μ_{y_j} (and in μ_{x_j}).

The step-down procedure tests sequentially

$$
H_0^{(1)} \qquad \mu_{y_1} = \mu_{x_1} \quad \text{versus} \quad \mu_{y_1} \neq \mu_{x_1} ;
$$

then

$$
H_0^{(2)} \qquad \mu_{y_2} = \mu_{x_2} \quad \text{versus} \quad \mu_{y_2} \neq \mu_{x_2} \quad \text{given} \quad \mu_{y_1}^* = \mu_{x_1}^* ,
$$

and so on, until

$$
H_0^{(q)} \qquad \mu_{y_q} = \mu_{x_q} \quad \text{versus} \quad \mu_{y_q} \neq \mu_{x_q} \quad \text{given} \quad \mu_{y_{q-1}}^* = \mu_{x_{q-1}}^* .
$$

The test statistic calls for the computation of a series of statistics T_j^2 defined as

$$T_j^2 = (\mathbf{Y} - \overline{\mathbf{X}})'_{(j)} S_{(j)}^{-1} (\mathbf{Y} - \overline{\mathbf{X}})_{(j)}, \quad j = 1, \ldots, q$$

which are the unconditional T_M^2-statistics for the first (preordered) q_j variables.

Now, for testing the j-th hypothesis we compute the statistics

$$G_j^2 = \frac{T_j - T_{j-1}}{1 + T_{j-1}^2/(n_2 - 1)}, \quad j = 1, \ldots, q$$

where $T_0^2 \equiv 0$.

The statistics which test the hypotheses in the step-down procedures are independently distributed under H_0 (Roy 1958, Subbaiah and Mudholkar, 1978). So, if we set up a control chart for each test statistic, the overall level of the procedure that simultaneously uses those charts can be easily obtained. This is a major advantage over both the procedures presented in Section 7.2 as well as over the procedures based on p univariate control charts. Another advantage over the p univariate charts is that in the test for each subhypothesis, the procedure uses the earlier variables as covariates, and so it properly accounts for the correlation structure.

Under the null hypothesis, the distribution of G_j^2 (conditional on $G_1^2, \ldots, G_{j-1}^2$) is

$$G_j^2 \sim \frac{(n_2 - 1)p_j}{n_2 - q_j} F(p_j, n_2 - q_j).$$

The q subhypothesis may be tested at different significance levels. Thus, the critical value for the j-th subhypothesis is

$$UCL_j = \frac{(n_2 - 1)p_j}{(n_2 - q_j)} F_{\alpha_j}(p_j, n_2 - q_j), \quad j = 1, \ldots, q$$

where α_j are not necessarily equal.

The step-down procedure does not reject $H_0 : \mu_Y = \mu_X$ if and only if $G_j \leq UCL_j$, for all $j = 1, \ldots, q$. The procedure can be considered as **an alternative** to the regular T_M^2 chart and not only as its supplement. The crux of the method is the plot of the q statistics $G_1^2, \ldots, G_q^2$ instead of the T_M^2-control chart.

If at least one of the G_j values exceeds the corresponding UCL_j, the process is declared statistically out of control, and a special cause of variation must be sought. The search should concern the subset corresponding to the first G_j^2 that signalled that the process is out of control. Note, that the control charts corresponding to later subsets in the a priori ordering lose their meaning since they test $\mu_{y_j} = \mu_{x_j}$ assuming $\mu_{y_{j-1}} = \mu_{x_{j-1}}$, which has been rejected.

We mentioned that the statistics G_j^2 are independently distributed. Therefore the overall level of the step-down procedure is

$$\alpha = 1 - \prod_{j=1}^{a}(a - \alpha_j) \, .$$

The step-down procedure was first formulated by Roy (1958) for the particular case in which the vector μ_y is partitioned into p subsets, each consisting of exactly one element. Subbaiah and Mudholkar (1978) developed the general case where each of the q subsets contains at least one element. For further references see Dempster (1963), Subbaiah and Mudholkar (1978b), Mudholkar and Subbaiah (1980a, 1980b, 1988), Kabe (1984), Marden and Perlman (1980, 1989, 1990), Wierda (1994).

Wierda (1994) analyzed the performances of the T_M^2- control chart, the step-down procedure, and the univariate procedure. He considers two performance characteristics: the power to detect a certain shift and the ability to indicate the type of shift. Based on a simulation study Wierda (1994) concludes that in the great majority of the cases the power of the step-down procedure is very close to that of the T_M^2-control chart. Furthermore, in general, the power of the univariate procedure is lower than those of the multivariate procedures. The exception is in the case when the shifts in mean are of a medium magnitude relative to the standard deviations, and the shift structure is in accordance to the correlation structure.

As for the ability to indicate the type of shift, the analysis has to be restricted to the step-down and the univariate procedures since the T_M^2-control chart provides no indication of that issue. The step-down procedure gives indication of the earliest among the q ordered subsets that has a mean shifted. Due to the conditional testing, nothing can be said about the means in later subsets on the ordering. The univariate control chart may provide extra information concerning the characteristics contained in those subsets.

For the step-down procedure, Wierda concluded that the highest probability of correctly detecting the characteristic (or subset) in which the shift occured, is obtained when the shift occurs in later subsets. In such cases, the step-down procedure takes advantage of the correlation structure, having the person's characteristics as covariates. As a result, he suggests that in practical settings, means which have a lower a priori probability of being shifted, should be included earlier in the orderings. Also, he suggests setting smaller α_j's for those subsets.

We performed a stepdown analysis on the tested ungrouped data (40 observations) from Case Study 1, as a two-step analysis: the first subset of variables contained the four diameter variables (i.e. variables 1-4); in the second step, we tested the significance of the contribution of the two length variables, conditional on the first subset. At $\alpha = .05$, the critical value for the G_j^2-variable is 13.05. We found that only in one case (case 49) the value was statistically significant. Note that in the case with the overall largest T_M^2 (case 66 with $T_M^2 = 82.63$) there was no contribution due to the length variables and all the deviations from standard originate from the diameter variables.

The test presented above is the likelihood ratio test for testing $H_0 : \mu_{y_j} = \mu_{x_j}$ versus $\mu_{y_j} \neq \mu_{x_j}$, given that $\mu_{y_{j-1}}^* = \mu_{x_{j-1}}^*$. Several other tests have been suggested, such as the tests based on $W_j = T_j^2 - T_{j-1}^2$ or on T_j^2, $j = 1, \ldots, q$. Basing the test on T_j^2 alone is not recommended since it ignores the information that $\mu_{y_{j-1}}^* = \mu_{x_{j-1}}^*$, and thus treats the variates and the covariates alike. Marden and Perlman (1990) compare the tests and conclude that for a large range of alternatives, the test based on G_j^2 has a higher power and is thus preferable.

The question on whether the assumption of an a priori ordering is too strong for practical settings is debatable. It can be argued that in many multivariate processes, not all quality characteristics are likely to show a shift in the mean, or at least not as likely as the others. The ordering and the levels of the step-down procedure are a quantification of those differences.

The step-down procedure cannot be used in cases in which there is a total lack of information on the a priori ordering. For such cases, Wierda recommends the case of the traditional T_M^2-control chart, supplemented by the univariate charts for detecting the responsible characteristics whenever T_M^2 signals for an out-of-control situation. He claims that "...for the time being there is no alternative." However, the statement, as such, is far from being accurate as we can find in the literature several multivariate methods which have been proposed for the detection of the out-of-control characteristics. However, the critical values suggested in the appropriate articles do not control the **overall** level of the test. We elaborate on this issue in the next section.

Other procedures for identifying out-of-control characteristics.

Mason, Tracy and Young (1995) present a unified approach for the procedures based on the decomposition of T_M^2. the procedure is presented here along with other methods which are briefly mentioned. The advantage of the approaches presented in this section is that no a priori ordering has to be assumed.

Let us start with the T_M^2 statistic

$$T_M^2 = (\mathbf{Y} - \overline{\mathbf{X}})' S_X^{-1} (\mathbf{Y} - \overline{\mathbf{X}})$$

and let us decompose it into independent components, each reflecting the (conditional) contribution of an individual variable. Assume that we wish to group the first $p - 1$ variables together and isolate the pth variable. Let us decompose the vectors $\mathbf{Y}, \overline{\mathbf{X}}$ and the matrix S, accordingly. The T_M^2-statistic based only on the first $p - 1$ variables is given by:

$$T_{p-1}^2 = (\mathbf{Y} - \overline{\mathbf{X}})'_{(p-1)} S_{X(p-1)}^{-1} (\mathbf{Y} - \overline{\mathbf{X}})_{(p-1)} .$$

Now, let us regress the pth variables on the first $(p - 1)$-variables in the n_2 observations of the base sample. Let us denote by $s_{p \cdot 1, \ldots, p-1}$ the residual standard deviation from the regression. The predicted value of the p-th variable of Y, given its values on the first $(p - 1)$ variables, is

$$\widehat{Y}_p = \overline{X}_p + \sum_{j=1}^{p-1} b_j(Y_j - \overline{X}_j) \,,$$

and its standardized value is

$$T_{p \cdot 1,2,\ldots,p-1} = \frac{Y_p - \widehat{Y}_p}{s_{p \cdot 1,\ldots,p-1}} \,.$$

It can be shown that

$$T_M^2 = T_{p-1}^2 + T_{p \cdot 1,2,\ldots,p-1}^2 \,.$$

Since T_{p-1}^2 is a regular T_M^2-statistic on $p - 1$ variables, we can further partition it into

$$T_{p-1}^2 = T_{p-2}^2 + T_{p-1 \cdot 1,\ldots,p-2}^2 \,,$$

and so on. Continuing to iterate and partition, we obtain

$$\begin{aligned} T_M^2 &= T_1^2 + T_{2 \cdot 1}^2 + T_{3 \cdot 1,2}^2 + \cdots + T_{p \cdot 1,2,\ldots,p-1}^2 \\ &= T_1^2 + \sum_{j=1}^{p-1} T_{j+1 \cdot 1,\ldots,j}^2 \,. \end{aligned}$$

The first variable, T_1^2 is the Hotelling T^2 for the first variable, which is the square of the univariate t-statistic,

$$T_1^2 = \frac{(Y_1 - \overline{X}_1)^2}{S_{X_1}^2} \,.$$

In the decomposition presented above there is an inherent ordering and as such, the assumption is not different in principle from that required in the step-down procedure. However, that decomposition is not the only one. There are $p!$ such decompositions and while the p terms within any particular decomposition are independent of each other, they are not independent across the different decompositions.

Under the appropriate null hypothesis, each of the terms is distributed as follows:

$$T_{j+1 \cdot 1,\ldots,j}^2 \sim \left(\frac{n_2 + 1}{n_2} \right) F_{(1, n_2 - 1)} \,.$$

Obviously, if one wants to ensure that the probability of at least one false rejection of the null hypothesis is α, the individual tests have to be performed at significance levels governed by appropriate multiple comparison procedures. Mason, Tracy and Young (1995) ignore the problem and suggest performing all the tests at the overall α-level. Given the large number of individual tests for any particular observation, we cannot agree with their recommendation. We suggest that it is important for the investigators to be aware of the possibility that if they perform all the tests at level α, the actual overall level of the test can be much larger than α.

As mentioned at the beginning of this section, other methods of decomposition have been suggested as well. The test proposed by Murphy (1987) for each particular subset of variables is based on the statistic

$$D_j = T_M^2 - T_j^2 \ .$$

This statistic contains a portion of the conditional terms in the decomposition of the T^2 statistic, since if we arrange the j variables as the first ones, we can write

$$D_j^2 = \sum_{\ell=j}^{p-1} T_{\ell+1 \cdot 1, \dots, \ell}^2 \ .$$

Hawkins (1991, 1993) defined a set of regression-adjusted variables in which he regresses each variable on all the others. His test statistics are the p-adjusted values, which can be shown to be related to the statistics presented in Mason, Tracy and Young's (1995) decompositions by:

$$z_j^{(1)} = \frac{T_{j \cdot 1, \dots, j-1, j+1, \dots, p}}{s_{j \cdot 1, \dots, j-1, j+1, \dots, p}} \ .$$

An additional approach presented by Hawkins is based on the decomposition of the T_j^2 statistics, using the standardized residuals from the j-th variable on the first $j-1$ variables, i.e

$$z_j^{(2)} = \frac{T_{j \cdot 1, \dots, j-1}}{s_{j \cdot 1, \dots, j-1}} \ .$$

8

The Statistical Tolerance Regions Approach

Objectives:

This chapter presents a non-standard approach to multivariate quality control that relies on the construction of statistical tolerance regions. The approach is particularly appealing to engineers for whom, in many cases, the concept of tolerance regions is very natural. The method is illustrated with Case Study 4 derived from an ongoing procedure to monitor the quality of citrus juice marketed as natural fruit juice.

Key Concepts
- Statistical tolerance region
- Level of confidence
- Proportion of multivariate observations
- Critical values
- Cross validation

A different approach can be taken to determine whether an observation, or a group of observations, come from a given multivariate normal distribution. Instead of setting up a "null hypothesis," one can identify a region which, under statistical control, contains a certain proportion of the multivariate observations. When the parameters μ and Σ of the distribution are known and the resulting region is called a normal process region, already mentioned in Chapter 1. Each new observation **Y** is tested to determine whether it belongs to the prespecified region by computing

$$T^2(\mathbf{Y}) = (\mathbf{Y} - \mu)'\Sigma^{-1}(\mathbf{Y} - \mu)$$

which is compared to the appropriate percentiles of the chi-squared distribution with p degrees of freedom. We illustrate the normal process regions with the data from Case Study 1. Assuming a multivariate normal distribution **with parameters equal to the empirical estimates from the reference sample of size 30**, we use the ungrouped estimate for the covariance matrix to compute the value of $T^2(\mathbf{Y})$ for the 66th observation whose measurements on the six dimensions are

$$(9.90, 9.89, 9.91, 14.88, 49.99, 60.14) .$$

The value of $T^2(\mathbf{Y}) = 82.63$ is larger than the 99.999-th percentile of the chi-squared distribution with 6 degrees of freedom. We thus conclude that the new observation does not belong to the mid 99.999% of the population defined by the "reference sample."

When the multivariate distribution is bivariate, the testing may also be performed graphically by plotting the new observation on the graph with the normal process regions. Figures 8.1 and 8.2 present the 50%, 90%, 95%, 99% and 99.73% bivariate regions corresponding (respectively) to the parameters of the two pairs of variables (**Diameter1, Length2**) and (**Length1, Length2**) on the first 30 observations. The regions were calculated under the same assumptions as in the calculation of $T^2(\mathbf{Y})$. On each plot we also marked the position of the 66-th observation with respect to the plotted variables. We observe that the tested observation does not belong to the central 99.7% of the bivariate distribution defined by the first pair of variables. (The same is true for all the 14 bivariate distributions whose pairs of variables contain at least one of

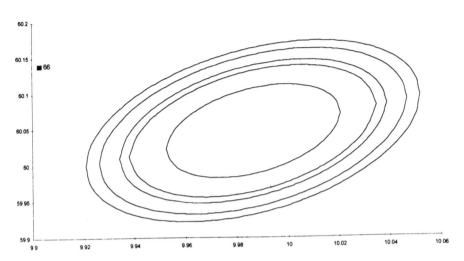

Figure 8.1: Bivariate normal density ellipses of **Length2** versus **Diameter 1** with parameters from the first 30 observations from Case Study 1 with 99.73%, 99%, 95% and 50% contours and observation 66.

the four diameters). However, on the second plot, the tested observation is well within the 90% normal process region.

As mentioned, the computation of the normal process regions requires knowledge of the parameters of the underlying distribution. When we cannot assume that the parameters are know (as is usually the case) one can still identify the distribution under statistical control using *Statistical Tolerance Regions* (Fuchs and Kenett, 1987, 1988). A region of observations is said to constitute a (P, δ) statistical tolerance region if we can assert with a level of confidence δ that the proportion of multivariate observations contained in that region is greater or equal to P. In one dimensional analysis, the statistical tolerance regions are called statistical tolerance intervals (ISO 3207, 1975) or statistical tolerance limits (ASQC, 1983). A new observation, **Y**, is tested to determine whether it belongs to a specific region by computing

$$T^2(\mathbf{Y}) = (\mathbf{Y} - \overline{\mathbf{X}})' S^{-1} (\mathbf{Y} - \overline{\mathbf{X}}) \,,$$

where $\overline{\mathbf{X}}$ and S are the empirical vector of means and the covariance matrix, respectively, as computed from the reference sample.

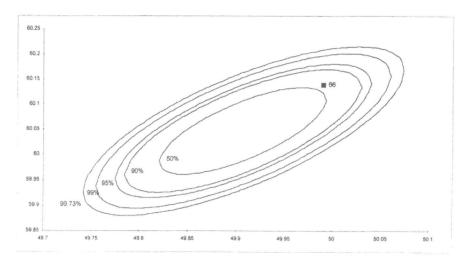

Figure 8.2: Length2 versus Length1.

When $T^2(\mathbf{Y}) \leq \kappa$, the confidence level in rejecting the statistical statement that $\mathbf{Y}$ belongs to the 100.p% central part of the distribution is given by δ. The practical implication of not rejecting the statistical statement is that the new observation is consistent with the process under statistical control. Fuchs and Kenett (1987) show that a good approximation of the critical value κ is given by

$$\kappa = \frac{g(P, p)(n-1)p}{g(1-\delta, (n-1)p)}$$

where $P_r(\chi_r^2 \leq g(\theta, r)) = \theta$, i.e. $g(\theta, r)$ is the lower θ^{th} percentile of a chi-square distribution with r degrees of freedom.

We illustrate the use of multivariate tolerance regions methodology in the context of quality control, with the data from Case Study 4 derived from an ongoing procedure to monitor the quality of citrus juice marketed as natural fruit juice. The data used below consist of a reference sample of pure juice ($n = 80$) and of an additional specimen to be tested. The attributes analyzed are labeled BX, AC, SUG, K, FORM, PECT (see Table 8.1). The means of the six attributes in the reference sample and in the tested specimen are presented in the first two rows of the table. Table

8.1 also contains the S the S^{-1} matrices computed from the reference sample. For example, the sample variance of BX (first element in the S matrix) is:

$$\sum_{h=1}^{80}(X_{1h} - \overline{X}_1)^2/(n-1) = 0.38$$

where $\overline{X}_1 = 10.4$ is the mean of BX in the reference sample. Also from the S matrix, the sample covariance between BX and AC is:

$$\sum_{h=1}^{80}(X_{1h} - \overline{X}_1)(X_{2h} - \overline{X}_2)/(n-1) = 0.035$$

and so on.

The quadratic form of Hotelling's T^2 which measures the distance between the values of the tested specimen (**Y**) and those of the reference sample is:

$$(\mathbf{Y} - \overline{\mathbf{X}})'S^{-1}(\mathbf{Y} - \overline{\mathbf{X}}) = 6.407 .$$

If we assumed for a moment that the parameters were known and equal to those found empirically in the reference sample, the value of $T^2(\mathbf{Y}) = 6.407$ would have corresponded to the 62.08-th percentile of χ_6^2, namely, the observation is not part of the 60% normal process region but it is contained in, say the 65% region. Let us now relax the assumption of known parameters and use the statistical tolerance regions methodology.

From the formula presented above for the approximation of the tolerance region for **Y**, we derived critical values for several values of P and δ with $n = 80$, $p = 6$. These are:

	$\delta = 0.7$	$\delta = 0.9$	$\delta = 0.95$
$P = 0.5$	5.539	5.825	5.969
$P = 0.6$	6.465	6.799	6.967
$P = 0.9$	11.003	11.571	11.858

By comparing the T^2 value of 6.407 to above critical values, we conclude that the tested specimen is not part of the central 50% of the

TABLE 8.1
Fruit juice example (Case Study 4)

	BX	AC	SUG	K	FORM	PECT
Y	11.9	1.3	8.9	1457.1	24.5	535.4
X̄	10.4	1.3	7.7	1180.0	22.2	451.0
Y − X̄	1.5	0.0	1.2	277.1	2.3	84.4
S-matrix						
BX	0.38					
AC	0.04	0.04				
SUG	0.29	0.01	0.26			
K	70.04	6.22	38.83	43590		
FORM	0.47	-0.22	0.38	207	6.53	
PECT	19.79	-1.93	12.05	5026	65.53	16554
S^{-1}-matrix						
BX	36.61					
AC	-25.12	60.39				
SUG	-35.01	20.96	38.31			
K	-0.02	0.00	0.02	0.00		
FORM	-0.67	2.45	0.35	-0.00	0.29	
PECT	-0.01	0.01	0.01	0.00	-0.00	0.00

reference population but is indeed contained in the 60% region. Obviously the specimen is also included in regions with coverage larger than 60%. The statement can be asserted for confidence levels above 0.7.

We note that we assumed the parameters to be known, the same $T^2(\mathbf{Y})$ did not correspond to a normal process region of 60%. The additional source of uncertainty created by the estimation of the parameters caused an expansion of the borders of the limits and the 60% statistical tolerance limits now include the new observation. Since the tested specimen was not found to be an extreme case of the reference population it is likely to be accepted as unadulterated. Alternatively, if the data analyzed above refer to a quality control test of an ongoing manufacturing

TABLE 8.2
Cross-validation results for juice example
(Table entries are tolerance limits κ and in brackets, the percentage of cases
with T^2 distance below the tolerance limits.)

			$100P$		
δ	50%	70%	80%	90%	95%
0.70	5.54	7.52	8.89	11.00	13.05
	(50.6%)	(67.9%)	(77.8%)	(86.4%)	(91.4%)
0.80	5.66	7.68	9.08	11.24	13.33
	(50.6%)	(67.9%)	(77.8%)	(87.7%)	(93.8%)
0.90	5.83	7.91	9.35	11.57	13.73
	(54.3%)	(67.9%)	(79.0%)	(88.9%)	(93.8%)
0.95	5.97	8.11	9.58	11.86	14.07
	(55.6%)	(71.6%)	(80.2%)	(91.4%)	(93.8%)

process, the results indicate that there is no reason to suspect that the process is out of control.

We note that unlike the normal process regions, the statistical tolerance regions are not as easily adapted to graphical representation and computer programs are not readily available even in the bivariate case. The decision making should thus rely on the analytical computations.

We proceed to evaluate the empirical performance of the tolerance region procedure under the null hypothesis, i.e. the hypothesis that the process is in control. Empirical assessments are necessary since specific data sets usually depart from the theoretical assumptions and the behavior of statistical methods may be sensitive to such departures. The performance of the tolerance region methodology will be evaluated subsequently by cross-validation. In the technique of cross-validation each observation is successively compared to the reference sample made up of the remaining $(n-1)$ observations. The percentage of observations inside the tolerance region is then estimated empirically by the proportion of cases (among n comparisons) that the tested observation was included in the tolerance region determined by the other $n-1$ observations.

The empirical results for our data set with $n = 80$ and $p = 6$ are summarized in Table 8.2.

We observe that the empirical percentage of coverage is close to the postulated P and quite insensitive to the value of δ. The method was thus found appropriate for routine testing of natural fruit juice authenticity. For more on the topic of multivariate tolerance regions and the citrus fruit juice case study, see Fuchs and Kenett (1987).

9

Multivariate Quality Control with Units in Batches

Objectives:

The chapter provides solutions to situations in which production is organized in batches and decisions are based on the results of samples from complete batches. Curtailing the inspection process is also illustrated with the objective to reach decisions faster and at lower costs. Indices that identify various sources of variability are defined and used to analyze real life data from a ceramic substrate manufacturer (Case Study 3).

Key Concepts
- Batch production
- Out of control process
- Corrective action
- Curtailing of inspection
- Overall variability decomposition

This chapter presents a quality control procedure for production batches which relies on the multivariate statistical tolerance region approach discussed in Chapter 8. Quality control of production lots typically relies on sampling and analyzing samples of units in order to determine if the production process is in statistical control. We consider here the situation in which the production is organized in batches and control is achieved at time intervals corresponding to the completion of entire batches.

When the production process is determined to be out of control an alarm signal is given to trigger corrective action. The quality control procedure appropriate for these conditions is illustrated with data from Case Study 3.

A set of identical modules are manufactured simultaneously on one ceramic substrate plate. The modules are separated before final packaging along lines inscribed on the substrate with a laser inscribing process. Exact positioning and dimensions of these lines determine, to a large extent, the success of the printing and component mounting operations. The ceramic substrates come in production batches of about 100 items. The quality control procedure used by the manufacturer is to draw a sample, whose size varies from 4 to 14, and carry out on these sampled units various measurements which are subsequently recorded. Because of the variety of modules manufactured, specifications of the substrate dimensions vary from batch to batch in nominal target values but not in specifications width. In this case study, we use deviations of the actual measurements from the center of the specifications as the analyzed observations. This normalization enables us to treat all batches on a uniform basis. The assumption made is that there are no location effects. A preliminary study validated this assumption.

All five dimensions (a, b, c) and (W, L) from Case Study 3 are considered. As mentioned in Chapter 3, the (a, b, c) measurements are determined by the laser inscribing process while the (W, L) are outer physical dimensions. These dimensions are measured on a different instrument than the first three. The two sets of measurements have in common the batch affiliation. The data presented in the Appendix on Case Study 3 detail the measurements for four (out of 31) such batches.

There are two possible outcomes to the decision process: 1) determine the production process to be in statistical control and, in practical terms, accept the batch, 2) stop the production flow as a result of a signal that the production process is out of control and start a 100% screening of substrates. During screening, we have to decide, for each item, if it meets our standard or if it has to be scrapped. We want to reach a decision (reject and stop testing entire batch) as soon as possible. The straightforward application of the multivariate quality control methods requires the testing of the entire sample. However, a complete test may be wasteful when results from the first few items tested suffices to stop the process. Therefore a curtailed decision scheme, that does not necessarily require all units to be tested, may be preferable in this case. From the "reference sample" we determine standard achievable dimensions including the correlation structure between the measured variables. A stopping rule for curtailing the testing procedure is presented below.

Let $Y_1, \ldots, Y_n$ be the tested sample batch. The curtailed testing technique is based on the sequential analysis of the n observations. Define $V_i = 1$ if the T^2 distance of the i-th tested observation is greater than κ and $V_i = 0$ otherwise. We assume that the observations are independent. We recall that the critical value κ given in Chapter 7 defines a region which covers a proportion P of the standard population (with confidence δ). For the i-th observation, let us define the null hypothesis that the process is in statistical control to state as:

$$Pr(V_i = 0) \geq P.$$

The curtailing of the inspection can obviously occur only at an observation g for which $V_g = 1$. Given that the inspection was not curtailed prior to the g-th observation, we define by $N(g)$ the total number of individual rejections of the null hypothesis up to and including, the g-th observation, i.e. $N(g) = \sum_{i=1}^{g} V_i$. To illustrate the notation, let us assume that we test sequentially 13 observations and at the tenth observation we have the second rejection of the null hypothesis, with the first rejection being at the second observation. Then we have $g = 10$, $v_2 = 1$, $v_{10} = 1$ and $v_j = 0$ for $j = 1, \ldots, 9$; $j \neq 2$. Also, $N(10) = 2$, but, say $N(9) = 1$ and so on. The number g, of observations up to the $N(g)$-th rejection, is compared to a critical value $R(N(g))$ defined as follows: For each number of rejections U (out of n), $R(U)^{+1}$ is the minimal number of observations allowed upto the U-th individual rejection, without rejecting the overall null hypothesis. Thus, if $g \leq R(N(g))$, the $N(g)$-th rejection "came too early" and we curtail the inspection and declare that the proportion of coverage is less than P. To continue the previous example, if for a proportion of coverage $P = .95$, the corresponding values for $U = 2$ is $R(U) = 7$, then the second rejection at the tenth observation, does not warrant the decision that the proportion of coverage is less than 95%.

Let α be the prespecified significance level for the overall null hypothesis. The $R(U)$'s are the minimal integer values such that under the null hypothesis

$$Pr\left(\sum_{i=1}^{R(U)} V_i \leq U\right) \geq \alpha .$$

For fixed $R(U)$, the random variable $\sum_{i=1}^{R(U)} V_i$ has a negative binomial distribution and for a given α, we can compute $R(U)$ from the inverse of the negative binomial distribution.

Although a table of $R(U)$ can be computed for each $1 \leq U \leq n$, in practice, we need only compute the function for $U \leq U_0 \leq n$, where U_0 is the minimal value for which

$$Pr\left(\sum_{i=1}^{n} V_i \leq U_0\right) \geq \alpha.$$

The value U_0 defines the minimal number of individual rejections (out of n) for which we reject the null hypothesis. We would thus certainly reject the null hypothesis if we observe more rejections and/or if the number of inspected items up to the U_0 rejection is less than n. Table 9.1 presents the $R(U)$ values for $n = 13$, $P = 0.95$, and $\alpha = 0.05, 0.01$, and 0.001.

We observe that for the three values of α, $U_0 = 3, 4$ and 5, respectively. If for example, we adopt $\alpha = 0.01$, we reject the null hypothesis at or before the third observation when $\sum V_i = 2$ up to that stage. We reject H_0 at or before the 9th observation if $\sum V_i = 3$ (provided that

TABLE 9.1
Critical $R(U)$ values for curtailed inspection

	$\alpha = 0.05$	$\alpha = 0.01$	$\alpha = 0.001$
$U = 2$	7	3	–
$U = 3$	13	9	4
$U = 4$	–	13	9
$U = 5$	–	–	13

we did not reject it already from the results of the first two or three observations). Finally, whenever we obtain the fourth value of $V = 1$, we curtail the testing process.

We now return to our example related to the analysis of batches of ceramic substrates. The decision process in that analysis has two possible outcomes: to accept the batch as is or reject the batch and start 100% screening of substrates. The need for this decision process comes from our lack of confidence in the stability of the substate manufacturing process. During screening, we have to decide for each item whether it meets our standard or has to be scrapped. In the sampling mode we want to reach a decision (accept or screen) as soon as possible. Therefore a curtailed inspection scheme should be used. The standard used in our analysis consists of measurements on a prototype batch of substrates which, when used in the subsequent manufacturing stages, proved to be of good quality in that, that later no problems related to physical dimension surfaced. From the "reference sample" we determined standard achievable dimensions including the correlation structure between the measured variables.

An extended data set from Case Study 3 included the results of the analyses performed on the Reference batch and on 30 other batches of ceramic substrate measurements. The results of testing the Reference and the other 30 batches of ceramic substrates are presented in Tables 9.2 and 9.3. Table 9.2 contains the group means of the (a,b,c) measurements and additional information related to the adherance of the sample to the "standard." Table 9.3 contains the same information for the (W,L) measurements. Based on the results on small sample approximations for κ, the Mahalanobis distances are compared to the computed κ-values. Any

TABLE 9.2

Analysis of the batches for (a, b, c)-measurements

SMPL	m	means			REJ	C.I	T_0^2	T_M^2	T_D^2	I1	I2
		a	b	c							
REF	13	-1	.6	.9	0	13	36.0	0.0	36.0	0.0	1.0
1	13	-2.7	.4	.4	3	2	468.2	172.0	296.2	0.37	0.63
2	13	2.5	-.4	.2	10	2	326.6	235.8	90.8	0.73	0.27
3	13	3.7	1.7	.5	10	2	489.3	373.2	116.1	0.76	0.24
4	13	0.0	-.2	-.1	1	13	89.2	68.8	20.4	0.77	0.23
5	13	-.1	.2	.3	0	13	59.3	34.3	25.0	0.57	0.43
6	16	-2.7	1.8	2.0	13	2	438.1	154.0	284.1	0.35	0.65
7	13	14.7	2.2	1.6	10	3	12045.1	3466.2	8578.9	0.29	0.71
8	13	.4	-3.2	-4.3	13	2	3784.4	1315.2	2469.6	0.35	0.65
9	14	.5	.4	1.1	0	14	68.1	32.2	35.9	0.47	0.53
10	12	.7	1.3	.3	2	12	177.4	72.0	105.4	0.41	0.59
11	13	1.2	2.2	1.0	6	2	133.8	113.5	20.3	0.84	0.16
12	12	-.2	1.6	0.6	6	5	345.6	51.4	294.3	0.15	0.85
13	13	-.7	-4.8	1.3	13	2	687.3	500.8	186.5	0.73	0.27
14	13	.3	1.4	.7	5	8	135.8	41.9	93.8	0.31	0.69
16	13	.7	3.0	1.8	10	2	197.3	164.7	32.6	0.83	0.17
17	13	.7	1.5	0.9	3	8	97.9	59.1	38.7	0.60	0.40
18	13	.5	1.1	0.2	1	13	77.4	66.6	10.8	0.86	0.14
19	13	1.7	-.6	-.1	10	13	277.1	185.6	91.4	0.67	0.33
20	13	0	.3	0	0	13	61.1	54.7	6.3	0.90	0.10
21	14	.6	1.6	1.6	0	14	89.4	62.7	26.6	0.70	0.30
22	6	-.3	.2	.3	0	6	20.5	11.8	8.7	0.58	0.42
23	5	0	0	0	0	5	22.8	22.8	0.0	1.00	0.00
24	4	-.5	0	0	0	4	16.9	9.2	7.7	0.54	0.46
25	13	0	.7	.7	0	13	29.8	18.7	11.1	0.63	0.37
26	4	.3	.5	.5	0	4	17.2	10.4	6.7	0.61	0.39
27	6	0	.8	.8	0	6	15.5	7.7	7.8	0.50	0.50
29	13	.2	-.2	0	2	13	89.0	72.0	17.0	0.81	0.19
30	4	.3	-12.3	-12.3	2	4	9913.6	2861.9	7071.7	0.29	0.71

observation with a T^2 value beyond the 95% tolerance region leads us to reject the production process. The tables also include information on the rejections (REJ), as well as the number of items (C.I) to be inspected if we adopted a curtailed inspection technique.

Eighteen batches were rejected on the basis of (a, b, c)-measurements; seventeen batches were rejected on the basis of (W, L)-measurements. All rejected batches are subjected to 100% screening. Only batches 20, 22-27 were determined as not statistically different form the reference batch, implying that for these batches the process is in control and that there is no need to screen the production batch.

We note from the results that in most cases the non-standard batches could have been detected in the very early stages of the testing process. In several of the groups in which there are considerable number of rejections, the first two tested items yielded T^2 distances larger than κ - and the testing process could thus have been curtailed.

We mention in this context that some of the batches deviate considerably from the reference samples. On the other hand, the 13 (W, L)-observations in batch 18 and the 5 (a, b, c)-observations in batch 23 were at their respective centers of specifications. The T_0^2-values in the two cases differ, however, since in the reference samples the means of the (W, L)-measurements were at their centers of specification while the means of the (a, b, c)-measurements deviate from the centers of specification.

The rejection strategy defined above refers to the analysis of the individual observations within the sample and can lead to a decision before the whole sample is tested on all dimensions.

Once testing is completed, we can proceed to analyze the decomposition of the overall variability computed from the entire sample batch (T_0^2) into the two measurements of variability: T_M^2 and T_D^2. The decomposition of T_0^2 can be perceived as a descriptive tool rather than as an exact testing hypotheses procedure. Based on the fact that T_0^2, T_M^2 and T_D^2 have asymptotic χ^2 distribution with np, p and $(n-1)p$ degrees of freedom, respectively (e.g. Jackson (1985)), we compute two indices I_1 and I_2 for detecting whether the overall variability is mainly due to the distances between means of the tested sample batch and those of the reference sample, or to variability within the tested sample batch. The indices are defined as normalized versions of the the two components of T_0^2. Since the expected value of a chi-squared variable equals its number of degrees of freedom we first divide T_M^2 and T_D^2 by p and $(n-1)p$, respectively, i.e.

$$I_1^* = T_M^2/p$$

TABLE 9.3
Analysis of the batches for the (W, L)-measurements

| SMPL | m | means | | REJ | C.I | T_0^2 | T_M^2 | T_D^2 | I1 | I2 |
		W	L							
REF	13	0	0	0	13	24.0	0.0	24.0	0.00	1.00
1	13	13.5	4.5	13	2	1082.1	821.9	260.2	0.76	0.24
2	13	13.8	0.6	13	2	1113.0	842.5	270.5	0.76	0.24
3	13	16.1	4.5	13	2	1300.9	1156.1	144.8	0.89	0.11
4	13	4.4	4.4	12	2	187.0	182.5	4.5	0.98	0.02
5	13	5.5	4.8	11	5	276.3	239.8	36.5	0.87	0.13
6	13	16.8	7.5	14	2	2666.3	1491.8	1174.5	0.56	0.44
7	14	10.3	0.3	12	3	1112.2	475.2	637.0	0.43	0.57
8	13	13.5	0.7	13	2	1325.3	803.1	522.2	0.61	0.39
9	13	3.2	3.2	6	3	113.6	96.1	17.5	0.85	0.15
10	8	0.6	0.6	0	8	4.5	1.1	3.4	0.24	0.76
11	5	0.4	1.2	0	5	12.8	3.9	8.9	0.31	0.69
12	13	11.3	-6.4	13	2	1529.3	1062.1	467.2	0.69	0.31
13	13	-4.0	15.8	13	2	2438.7	2095.8	343.9	0.86	0.14
14	13	9.2	-5.6	12	2	1079.4	749.4	330.0	0.69	0.31
15	13	4.1	3.7	8	2	181.7	141.6	40.1	0.78	0.22
16	13	0.7	0.0	0	13	10.1	2.2	7.9	0.22	0.78
17	13	9.1	-1.5	12	3	498.6	422.8	75.8	0.85	0.15
18	13	0.0	0.0	0	13	0.0	0.0	0.0	-	-
19	13	11.1	3.1	11	2	1426.6	550.2	876.4	0.39	0.61
20	13	-0.3	-0.7	0	13	43.1	3.5	39.6	0.08	0.92
21	13	0.1	1.2	1	13	31.4	11.1	20.3	0.35	0.65
22	13	-2.0	-2.6	0	13	62.1	56.0	6.1	0.90	0.10
23	13	1.1	1.2	0	13	17.4	12.6	4.8	0.72	0.28
24	13	-0.2	0.5	0	13	13.7	2.7	11.8	0.20	0.80
25	13	0.9	0.8	0	13	14.5	7.3	6.8	0.53	0.47
26	13	1.1	0.8	0	13	13.8	8.3	5.5	0.60	0.39
27	6	1.5	1.3	0	6	10.4	8.5	1.9	0.82	0.18
28	13	-35.9	-43.	12	2	38288.2	15025.8	23262.4	0.39	0.61
29	13	-0.2	-0.	0	13	9.0	1.1	7.9	0.12	0.88
30	13	-0.1	-1.1	0	13	25.2	8.4	16.8	0.33	0.67

$$I_2^* = T_D^2/(n-1)p$$

and then divide those values by their sum, thus obtaining the indices I_1 and I_2 which sum to 1,

$$I_1 = I_1^*/(I_1^* + I_2^*)$$

and

$$I_2 = I_2^*/(I_1^* + I_2^*) \, .$$

It can be shown that in terms of the original T^2-statistics we have

$$I_1 = (n-1)T_M^2/\left[(n-1)T_M^2 + T_D^2\right]$$

and

$$I_2 = T_D^2/\left[(n-1)T_M^2 + T_D^2\right] \, .$$

When no source of variability significantly supersedes the other, the two indices should be approximately of the same order of magnitude. The normalized I_1 and I_2 are an extended version of the indices proposed by Fuchs and Kenett (1988). The indices I_1 and I_2 for the data in our example are presented in Tables 9.2 and 9.3.

The indices I_1 and I_2 are helpful in identifying sources of variability. We illustrate the assertion with two examples based on the (W, L)-measurements. In batch 4 we conclude from the very low value of I_2 (0.001) that there is almost no variability within the sample. The large number of rejections (12 out of 13) indicates an almost constant bias of the magnitude of the sample means with respect to the reference sample (and, in this case, also with respect to the centers of specification). On the other hand, in batch 20, a significant part of the variability is due to the variability within the sample ($I_2 = .49$), with the (W, L)-means being close to those in the reference sample.

The variability of the (W, L)-measurements in batch 20 did not suffice to yield rejections of the individual substrates. In other cases, such as in the (a, b, c)-measurements in batch 12, the deviations were large enough to generate six rejections. Further analyses in a similar manner can be performed on the indices of the other samples. We note that several measurements differed so drastically from the standard, that errors in recording are to be suspected. The 13 (W, L)- observations in batch 28 yielded a very large T_0^2-value due to large deviations from the standard. The substrates were found to be smooth and not inscribed and thus no (a, b, c)-measurements are available. The batch has obviously been rejected. Data on the (a, b, c)-measurements were also not available for batch 15.

10

Applications of Principal Components

Objectives:

The chapter provides procedures for constructing principal components with an emphasis on their interpretation in multivariate quality control applications. In some cases, principal components have proven to be an effective approach in reducing dimensions of multivariate data. The aluminum parts data (Case Study 1) is used to illustrate the approach.

Key Concepts
- Eigenvalues
- Eigenvectors
- Principal component charts

A quite common approach in multivariate quality control is to attempt to reduce the dimensionality of the data by transforming the original p variables into a lower dimensional data set obtained by identifying several meaningful linear combinations of the p dimensions. The construction of specific linear combinations, called *principal components*, is described here. Case Study 1 will be used to illustrated in the application of principal components in multivariate quality control.

Let $\mathbf{X} = (X_i^{(1)}, X_i^{(2)}, \ldots, X_i^{(p)})$ be the i-th p-dimensional observation in the original data. We create a new p-dimensional observation $(Z_i^{(1)}, \ldots, Z_i^{(p)})$, such that the ℓ-th variable in Z, $Z_i^{(\ell)}$ is a linear combinations of the deviations of the original p dimensions from their targets, i.e.:

$$Z_i^{(\ell)} = \sum_{u=1}^{p} c_{\ell u}(X_i^{(u)} - m^{(u)}) .$$

Let us now present in more detail the rationale and the construction of the principal components. For a process with multidimensional data, its overall variance is defined as the sum of the variances of the p-variables, i.e. the sum of the diagonal elements of the covariance matrix $\Sigma = [(\sigma_{\ell u})]$. If we denote by $\theta_1, \theta_2, \ldots, \theta_p$ the characteristic roots (eigenvalues) of the covariance matrix Σ, it is well known that their sum equals the overall variance of the process defined above, i.e. $\Sigma\theta_\ell = \Sigma\sigma_{\ell\ell}$.

Furthermore, if we order the characteristic roots such that $\theta_1 \geq \theta_2 \geq \cdots \geq \theta_p$, the matrix C whose rows are the corresponding characteristic vectors (eigenvectors) $C_1, \ldots, C_p$ satisfies $CC' = \Sigma$, $C'C = \Omega$ where Ω is the matrix having the θ's on the diagonal and zero for all the off-diagonal elements. Now it can be shown that if we make the linear transformations

$$Z^{(1)} = C_1'(\mathbf{X}-\mathbf{m}); \quad Z^{(2)} = C_2'(\mathbf{X}-\mathbf{m}); \quad \cdots; \quad Z^{(p)} = C_p'(\mathbf{X}-\mathbf{m});$$

the new $Z_i^{(\ell)}$ variables will be uncorrelated, and if the covariance matrix Σ is known their respective variances are θ_ℓ, $\ell = 1, \ldots, p$. For the i-th observation, the linear combinations $Z^{(1)} = C_1'(\mathbf{X} - \mathbf{m})$ has the maximal variance among all possible linear combinations of the X's. The proportion $\theta_1 / \Sigma\theta_\ell$ is sometimes termed the "variance explained" by the first principal component. Similarly, for $\theta_2 / \Sigma \theta_\ell$ and so on. Parsimony is achieved if for the several last principal components the ratio $\theta_u / \Sigma \theta_\ell$ is small enough to justify ignoring the corresponding Z-variables from the analysis.

When the covariance matrix Σ is not known, but rather estimated, the Z-variables will still be orthogonal, and Ω is now their estimated rather than exact covariance matrix.

Analyzing Data with Principal Component Charts.

For each multivariate observation, the principal components are linear combinations of the deviations of the p variables from their respective targets. Principal components have two important features which are useful in quality control:

(a) the new variables are uncorrelated (or nearly so), and

(b) in many cases, a few principal components may capture most of the variability in the data so that we do not need to use all the p principal components for control.

Let us illustrate the feature of parsimony mentioned in (b) above with the data from Case Study 1. The first column of Table 10.1 contains the eigenvalues of the inverse of the covariance matrix for the first 30 ungrouped data points. We notice that no more than three out of the six principal components are worth analyzing. The first principal component explains 67% of the overall variation, the next one is responsible for additional 21.5% and the third explains another 10%. The variance accountable for the last three components is ignorable (less than 2%).

The advantages of principal components charts over individual Shewhart control charts (uncorrelated variables and possible parsimony) are obtained by losing the original identity of the variables. However, in some cases, the specific linear combinations corresponding to the principal components with the largest eigenvalues may yield meaningful measurement units. In our example, the columns labeled Vec1 to Vec6 of Table 6 display the eigenvectors corresponding to the six eigenvalues presented in the first column. These are the coefficients of the original variables used in the definition of the principal components. The first three principal components, which explain almost all the variance, are

TABLE 10.1
Eigenvalues and eigenvectors of covariance matrix
first 30 observations of case study 1

	E-Values	%	Cumul %	Vec 1	Vec 2	Vec 3	Vec 4	Vec 5	Vec 6
1	240E-03	67.02	67.02	-0.04	0.46	0.11	-0.26	-0.79	0.29
2	7.70E-04	21.47	88.49	-0.05	0.46	0.13	-0.26	0.07	-0.84
3	3.59E-04	10.00	98.49	-0.04	0.50	0.10	-0.40	0.60	0.46
4	2.86E-05	0.80	99.29	-0.03	0.51	0.16	0.84	0.06	0.05
5	1.45E-05	0.40	99.69	0.70	-0.12	0.70	-0.03	0.01	0.00
6	1.10E-05	0.31	100.00	0.71	0.23	-0.67	0.01	-0.01	-0.02

clearly interpretable in terms of the physical dimensions analyzed. The first component is defined almost exactly as an average of variables 5 and 6, i.e. the two length variables. The second component is basically an average of the other four variables which are diameter measurements at various locations on the pin, although the contribution of the length measurements is not totally ignorable. Finally, the third component is interpretable as the difference between variables 6 and 5, which is in our case the length of the main part of the pin (without the cap).

Since the principal components are linear combinations of the deviations of the variables from the targets, when the process is under control their values should be around zero with variances determined by the eigenvalues. Thus, if we divide all the principal components by their respective standard deviations, (i.e. by the square roots of the respective eigenvalues) we can plot the resulting standardized principal components on charts with mean value set at zero and constant UCL and LCL of ±3 corresponding to 3-standard deviations control limits. Each principal component can be plotted on a separate chart and pairs can be plotted on a scatter plot of two (usually the first two) components.

Figure 10.1a and Figure 10.1b present the principal components charts for the first two principal components and Figure 10.2 displays the corresponding scatter plot. We observe that the first principal component has significant negative value at the 48-th observation and significant positive value for observations 49 and 61. The second principal component has a significant outlier at the 66-th observation. Given the

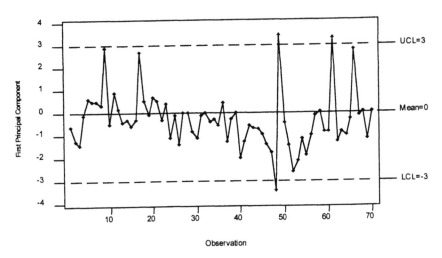

Figure 10.1a: Principal components chart for the first component of data from Case Study 1 (ungrouped).

coefficients of those components, those outlying values should correspond to outliers in the length and diameter measurements, respectively. We shall confirm this observation later in Chapter 11 where we present methods for simultaneous display of univariate and multivariate statistics.

Meaningful interpretation of the first principal components are quite common in practice and have been reported in the literature. In a frequently quoted example of ballistic missile data (Jackson 1959, 1985), the coefficients in the first principal component had all the same sign and were roughly of the same magnitude. The resulting variable was thus related to the average of the original variables and could be considered to measure the product variability. Further, as in the ballistic missile example, when the coefficients for the next one (or more) principal components differ in signs and are pairwise of the same order of magnitude, the resulting principal components represent differences between the means of the respective subsets of variables.

Obviously, the coefficients of the variables in the principal components do not have to obey the quite simplistic patterns mentioned above. Sometimes, as in our example, the coefficients corresponding to several

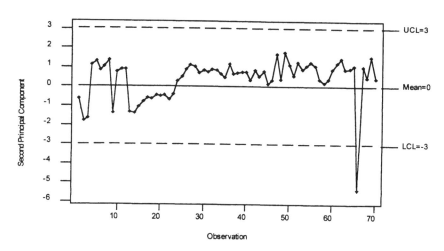

Figure 10.1b: Principal components chart for the second component of data from Case Study 1 (ungrouped).

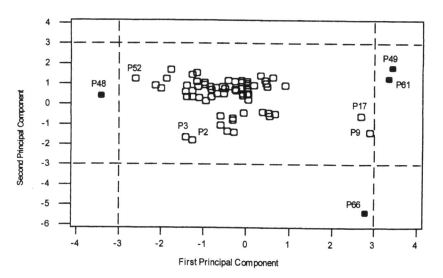

Figure 10.2: Scatterplot of the first two principal components of data from Case Study 1 (ungrouped).

attributes are close to zero. The resulting principal components can then be considered to represent linear combinations of only a subset of the p variables. Moreover, when the coefficients for the various attributes differ in their order of magnitude, other types of interpretations (if at all) can be attached to the resulting principal components, such as the comparison of one variable with the average of several others and so on.

Since the principal components do not provide an overall measure of departure of the multivariate data from the norms, those charts may complement but not replace the multivariate T^2-charts presented in the previous section. The analysis of the several charts simultaneously suffer from the problems mentioned in connection with the assessment of an out-of-control situation from the control charts of the original variables (although the number of relevant charts is likely to be somewhat reduced in the principal components analysis). Moreover, as in the T_M^2-chart, the individual attributes most prominently responsible for an out-of-control observation are unidentifiable.

11

Additional Graphical Techniques for Multivariate Quality Control

Objectives:

The chapter provides additional graphical techniques using computer technology for analyzing multivariate data in quality control applications. The star plots and the MP-charts are particularly emphasized.

Key Concepts
- Computerized graphical displays
- Symbol plots
- Three dimensions plots
- Scatterplot matrix
- Start plots
- MP-charts

Background on Modern Computerized Graphical Displays

The display of data is a topic of substantial contemporary interest and one that has occupied the thoughts of many scholars for almost 200 years. A pioneer in this area was W. Playfair who in his 1786 *Commercial and Political Atlas* displayed differences between English imports and exports in informative and resourceful ways (see e.g. Costigan-Eaves and MacDonalds-Ross, 1990). A vision for data analysis in the 80's was outlined by John Tukey (1962), who has reiterated a perspective focused on data-based graphics (Tukey, 1986). Tukey's vision has largely materialized in the Exploratory Data Analysis approach whose foundations was established in the books by Tukey (1977), Mosteller and Tukey (1977), and Velleman and Hoaglin (1981).

Modern computerized data analysis builds on advances in computer technology and statistical methodology. Computing environments specifically designed for data analysis have evolved since the 60's on computing platforms that reflect the state of the art in computing technology: first mainframes, then personal computers, and now workstations. Such environments incorporate aspects of data analysis that include model identification, diagnostic checking, transformation and the usual inferential procedures (Thisted, 1986; Huber, 1987). Becker et al. (1987) specify requirements for computerized dynamic graphics methods that include identification, deletion, linking, brushing, scaling, and rotation. These techniques are briefly defined below and some of them are then illustrated in Figures 11.1a-11.1c with data from Case Study 2.

1. Identification: Labelling and locating of specific observations on graphical displays.

2. Deletion: Deleting specific points from a graph and redrawing the graph.

3. Linking: Linking methods that make effective use of color and/or plotting symbols for indicating the value of a dimension not explicitly present in the graphical display.

4. Brushing: Conditioning the analysis on the different values of any variable.

5. Scaling: The ability to refocus, move and resize graphs.

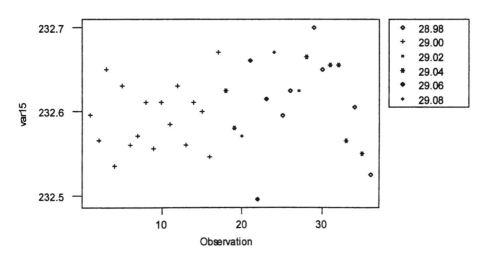

Figure 11.1a: Time sequence plot of variable 15 from Case Study 2 with symbols indicating levels of variable 14.

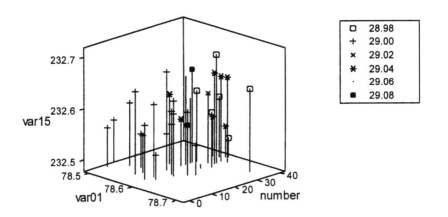

Figure 11.1b: 3D-Plot of variable 15 from Case Study 2 with symbols indicating levels of variable 14.

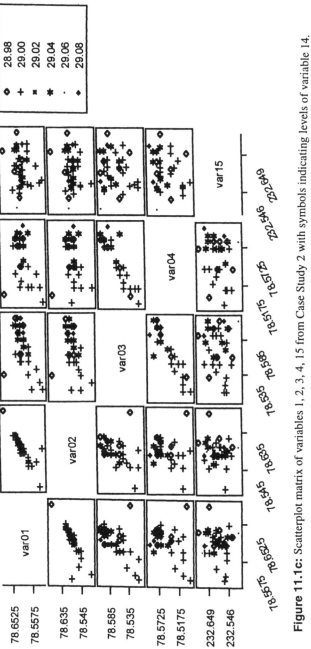

Figure 11.1c: Scatterplot matrix of variables 1, 2, 3, 4, 15 from Case Study 2 with symbols indicating levels of variable 14.

6. Rotation: The ability to rotate three dimensional scatterplots in either an automatic continuous motion mode or in a manual mode.

Several software packages for personal computers, that include such features, are now available (e.g. MINITAB™, EXECUSTAT®, JMP®, DataDesk® and STATGRAPHICS®).

Case Study 2 consists of data collected in a process capability study of a complex mechanical part on which 11 diameters, 4 length and 2 wall-width measurements were recorded. In our analysis we focus on the overall length of the part (Variable 15) and use MINITAB for the analysis. Figure 11.1a presents a time sequence plot of Variable 15 using symbols to indicate the levels of Variable 14, an internal measurement of the part. The shortest part is the 22nd (232.495 mm). Its value on Variable 14 (29.06 mm) is one of the largest in the sample. However, such large values of Variable 14 occurred also on longer parts. Figure 11.1b presents a three dimensional scatterplot of Variable 15 versus the observation index and versus Variable 1. Again, the points are labelled with symbols indicating the level of Variable 14. The figure indicates that the group of the first 10 parts are low on both Variable 1 and Variable 14. We observe an outlying point with a very low value for Variable 14, and an unusually high value for Variable 1. Figure 11.1c is a matrix of scatterplots for Variables 1, 2, 3, 4 and 15, again, with the labelling of the observations to indicate the level of Variable 14. The two diameters - Variable 1 and Variable 2 - are clearly linearly related. The same type of relationship appears to hold between Variable 3 and Variable 4. It is also worth noting that the symbols for Variable 14 tend to cluster in the Variable 2 versus Variable 3 scatterplot, thus indicating a further correlation with Variable 14.

The graphical displays presented above provide the reader with a glimpse at the capabilities of modern interactive data analysis. Indeed, computerized data analysis has become a new discipline, not merely an extension of Playfair's graphical innovations. Industrial statisticians are now developing comprehensive approaches to computerized data analysis. Comprehensive example of a strategy for on line multivariate exploratory graphical analysis was developed by the CIBA-GEIGY Mathematical Applications group in Basel (Weihs and Schmidi, 1990). For another example of dynamic graphics for exploratory analysis of spatial data using a "map view" see Haslett et al., 1991. For an algorithmic approach to multivariate quality control see Chua and Montgomery, 1991.

Graphical Techniques for Quality Characteristics in Multivariate Quality Control.

In the previous chapters, we mentioned that when a T^2-value or the value of a principal component corresponding to an observation (or to a group of observations) exceeds the critical limit marked on the graph, one cannot identify the attributes primarily responsible for the overall outlying value. This has been illustrated in our example where several observations (or subgroups) were identified as outliers, but without the indication on the behavior of the particular characteristics. Additionally, unlike the univariate case, the scale of the values displayed on the chart is not related to the scales of any of the monitored variables. Both pieces of information are relevant and important when one monitors the process. On the other hand, we have already described the shortcomings of relying only on univariate charts when the data are intrinsically multivariate. Thus, in the multivariate case, neither the univariate charts for each variable, nor the multivariate chart provide by themselves a complete picture of the process.

The problems raised above can be considerably alleviated if all the information about the individual characteristics as well as the overall measure of distance from targets are displayed on the same chart. Several graphical methods for coping with these problems have recently been proposed. Two such methods will be presented in this chapter: the starplot (STATGRAPHICS 3.0, 1988) and the MP chart (Fuchs and Benjamini 1991). These plotting techniques are illustrated with the six dimensional data from the process capability study for turning aluminium pins (Case Study 1). As in most of the previous section we start by analyzing the data grouped in pairs. Along the presentation of the graphical display we will return and address some of the findings mentioned in chapters 5 and 9, thus complementing the analyses performed with multivariate quality control and principal component charts.

The quality characteristics which are to be displayed simultaneously on the charts are on different scales. Therefore, proper scaling of the data is required. Furthermore, if we want to relate each variable to its reference value, the statistics displayed have to be (scaled) deviations from those targets. Let us define for the ℓ-th variable in the j-th subgroup of size n, the standardized deviation of $\overline{X}_j^{(\ell)}$,

$$d_j^{(\ell)} = \frac{\overline{X}_j^{(\ell)} - m^{(\ell)}}{v_\ell}$$

where $X_{ij}^{(\ell)}$ is the i-th observation in subgroup j on the ℓ-th variable, $\overline{X}_j^{(\ell)} = \sum X_{ij}^{\ell}/n$, $m^{(\ell)}$ is the target value and v_ℓ is the scale factor for the ℓ-th variable, estimated from the base sample. As an alternative, one can consider the scaled data rather than the scaled deviations, i.e.

$$h_j^{(\ell)} = \frac{\overline{X}_j^{(\ell)}}{v_\ell}.$$

The multivariate measure of distance from target displayed on the three types of charts is the T^2-statistics. When the data are grouped, we can also construct simultaneous charts for dispersion. In those cases, the multivariate measures of dispersion displayed are the values of T_D^2 as presented in the previous section. On the T^2-charts, we can also display some common univariate statistics (i.e. the ranges or standard deviations), again after proper scaling. The simultaneous charts for dispersion have been developed only for the MP-charts but obviously they can be constructed as companions to the starplot.

The multivariate charts presented here are symbolic scatter plots. The starplot for groups is shown in Figure 11.2 and the MP-charts is displayed in Figure 11.3. Theoretical considerations and empirical evidence showed the MP-chart to press desirable properties (Fuchs and Benjamini, 1991). The starplot has already gained relative popularity and has been implemented in other software packages (e.g. SAS).

The stars in the starplot and the symbols in the MP-chart are positioned vertically at the group T^2-value. Thus, those charts resemble the multivariate Shewhart charts, except that for each group we display at its T^2-value a more complex symbol than a single dot. We describe below the two types of charts and illustrate their use with the data from Case Study 1.

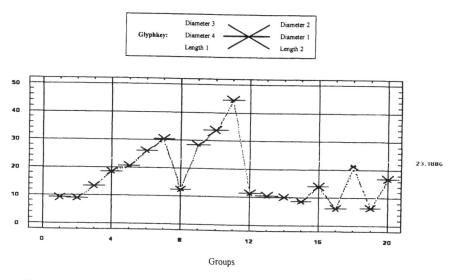

Figure 11.2: Starplot of the 6 variables from Case Study 1.

Starplots

The starplot was apparently first developed at the SCS corporation as an enhancement to the Multivariate Control Charts available through the STATGRAPHICS statistical software package. The starplots consist of stars positioned vertically at the group T^2-value (see Figure 11.2).

Each multivariate observation is portrayed in the starplot by a star with the rays proportional to the deviations of the individual variables from the minimal values across all observations (or groups of observations). Thus, the effect of displaying the $h_j^{(\ell)}$-values rather than the $d_j^{(\ell)}$'s is that when when the process is under control the stars are of medium size and when a variable is out of control with a value considerably below the target, the ray shrinks to a minimal value, attracting minimal visual attention. Moreover, that starplots give no indication of the statistical significance of the ray's length.

MP–charts.

The MP–charts were proposed by Fuchs and Benjamini (1991) as an improvement over the starplot and other methods proposed in the literature.

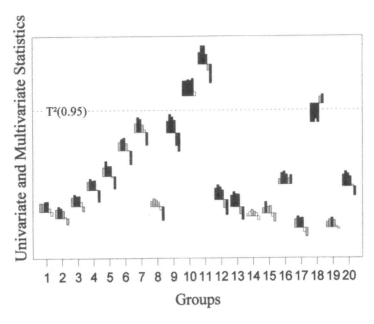

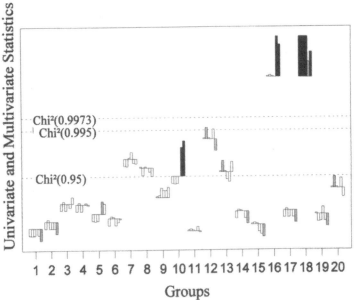

Figure 11.3: Mp-charts for means and dispersion for the 20 tested groups from Case Study 1.

The symbol in the MP-chart is an adaptation of a profile plot which encodes visually the size and deviation of the group average of each variable from its target. After the horizontal base line of a symbol is drawn, a bar is constructed sequentially for each variable. The bar extends either above or below the base line according to the sign of the deviation. The bars are proportional to the absolute value of $d_j^{(\ell)}$ up to a value of 4, after which they remain at a constant length of 4. The bar darkens (or changes color) as the level of significance of the deviations increases (see Figure 11.3).

Minor adjustments are necessary on the vertical axis to account for the size of the symbol. The critical value corresponding to a T^2-value of .997 (equivalent to the 3 standard deviations rule in the univariate chart) is at one symbol-size distance from the top, and that corresponding to 0 is placed half a symbol size distance from the bottom. As mentioned, the MP-chart for dispersion is constructed along the same lines as the MP-chart for means and together they enable the investigator to study both the level and the variation in the process.

Reexamination of Case Study 1.

We conclude this chapter with a renewed analysis of Case Study 1. The analysis provides a comparison of graphical methods as well as an extension to the analysis derived earlier by the use of multivariate control charts and principal components.

Let us now reiterate the main findings from the aluminium pins process capability study mentioned in the analysis of the Shewhart charts, the T^2-charts and the principal components:

(a) From the Shewhart charts on the sixth variable (**Length2**), we observed that the overall length is significantly below target in groups 24 and 26 and that that there are large dispersions between the two observation which form group 25 and 31 (Figure 6.1).

(b) In the analysis of the T_M^2-charts we mentioned that at the 26-th subgroup the deviations from the target are large enough (at least on some variables) such that the overall T_M^2-value exceeds the critical 0.997 value. Also, we noticed that the T_D^2-chart shows evidence of large dispersions between the two observations which form the subgroups 31 and 33 and, in a lesser extent, subgroup 27 (Figures 6.2a and 6.2b).

(c) The assessment of the principal components (for ungrouped data) revealed that the first principal component had a significant negative value at the 48-th observation and significant positive value for observations 49 and 61. The second principal component had a significant outlier at the 66-th observation (Figures 10.1a and 10.1b). We mentioned that the the coefficients of the first two components are such that the components roughly correspond to the mean of the two length measurements and the mean of the four diameter measurements, respectively.

The Shewhart control chart in Figure 6.1 flags subgroups 24, 26, and 27 as having a significantly low value for variable 6. Only subgroup 26 stands up as significantly different on the T^2-chart. The starplot indicates only implicitly the decrease in variable 6 from subgroup 10 to subgroup 24 and that the measurements are significantly below target. See for example subgroup 33 in Figure 11.2. As mentioned above, another drawback of starplots is that the rays carry no indication about the significance level of the departures from targets.

We now turn to examine Figure 11.3 which contains the MP-charts for means and dispersion for the last 20 tested subgroups. First we mention that the findings from the Shewhart charts about the sixth variable (**Length2**), are also directly observable from the MP-charts. Secondly, about the findings from the T^2-charts, we observe that indeed in the 26-th subgroup (i.e. the 11-th tested subgroup) the value of the sixth variable is significantly below its target. The second and the third variables exceed their respective targets by 2 to 3 standard deviations. On the other hand, subgroup 33 (i.e. the 18-th tested subgroup) has three highly significant deviations from target (at variables 1, 2, and 4) but its overall T^2-value is smaller than in group 26. The apparent inconsistency is easily explained, as it it clearly seen from the graph, by checking the correlations among the variables. In group 26 (the 11-th tested subgroup) a very large deviation in the sixth variable is accompanied by a much smaller deviation in the fifth one. This pattern does not correspond to the the high positive correlation between those variables in the base sample. On the other hand, the pattern in subgroup 33 (the 18-th tested subgroup) is as expected, i.e., the four diameter measurements deviate together (in this case considerably below their targets) and the two length measurements

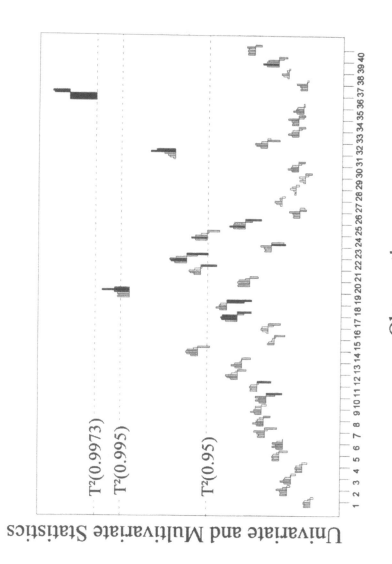

Figure 11.4: MP Chart for the tested observations of Case Study 1 (Ungrouped).

are again together (less than 2 standard deviations above their respective targets).

The large within group dispersions which yielded the large T_D^2 values for groups 31 and 33 (the 16-th and the 18-th tested subgroups, respectively), are caused by two different sets of variables. We observe from the MP-chart for dispersion that in group 31 the two length measurements vary between the observations (61 and 62) which form group 31. In group 33 the diameter measurements are responsible for the large T_D^2-value. Two interesting patterns are observed for groups 25 and 27 (the 10-th and the 12-th tested subgroups, respectively). In group 27, although none of the within group variances are mentionable, the T_D^2-value is large. The reason lies in a quite intricate configuration of correlations among variables 1–4 in the two observations 53 and 54 which form that group. On the other hand, in group 25 although large within group variances are observed for the length measurements, the overall T_D^2-value is quite small. This is due to the unusual small variances in the first three variables, and to the adherence to the postulated correlation structure.

The pattern of the variables for the observations within the subgroups are best observed from the ungrouped MP-chart for means. Figure 11.5 presents that chart for the 40 tested observations. From that chart we can also reexamine the findings from the principal components chart. The significant negative value of the first principal component at the 48-th observation (i.e. the 18-th tested observation) is immediately related to the significant negative deviations in the two last variables which are the length measurements. Similarly, significant positive deviations in those variables in observations 49 and 61 (i.e. the 19-th and the 31-st tested observations) yielded significant value in the first principal component. The interesting pattern in observation 66, (i.e. the 36-th tested observation) where the diameter measurements (first four variables) are significantly below their targets, while both of the length measurements exceed by 2 to 3 standard deviations their respective targets, are reflected in the significant negative value of the first principal component and in the borderly significant value of the second principal component.

12

Implementing Multivariate Quality Control

Objectives:

The chapter provides a strategy and a generic plan for implementing the multivariate quality control methods presented in the book. The proposed plan specifies steps and milestones that have been proved to be effective in successful implementation of quality control.

Key Concepts
- Quality control process
- Control subjects
- Common causes
- Chronic problems
- Special causes
- Sporadic spikes
- Strategic plan for implementing Multivariate Quality Control

Maintaining control is a basic activity that can be observed wherever there is change. For example, the human body exerts control in its efforts to maintain body temperature, blood count, and other physiological functions. Industrially, it takes the form of meeting goals: delivery according to schedule, using funds according to budget, and producing quality according to specification.

In this chapter we discuss the basic elements of multivariate quality control and then present a planning strategy for developing such a quality control process. The chapter provides to its reader an integrated review of the topics presented in the above nine chapters. Juran (1988) specifies a series of 7 steps which, when applied to problems of quality, effectively implement a quality control process. These steps create a feedback loop which produces quality control. First, a goal, target or standard is established as the intended value of the control subject. Then, when it is determined that the difference between actual performance and target warrants an operator intervention a control response is being activated.

In multivariate quality control these steps require special consideration. We elaborate on these special issues by commenting on Juran's 7 steps in the context of multivariate quality control. We label the steps S1-S7. A flowchart of these steps is presented in Figure 12.1.

S1. Determination of the control subjects to be used for quality control.

In general, a production process has many sources or causes of variation. Complex interactions between material, tool, machine, work methods, operators, and the environment combine to create variability in the process. Factors that are permanent, as a natural part of the process, cause chronic problems and are called common causes of variation. The combined effect of common causes can be described using probability distributions. The only way to reduce the negative effects of chronic, common causes of variability, is to modify the process. Special causes, assignable causes or sporadic spikes arise from external temporary sources that are not inherent to the process. In order to signal the occurrence of special causes we need a central mechanism. The statistical approach to process control allows us to distinguish between chronic problems and sporadic spikes.

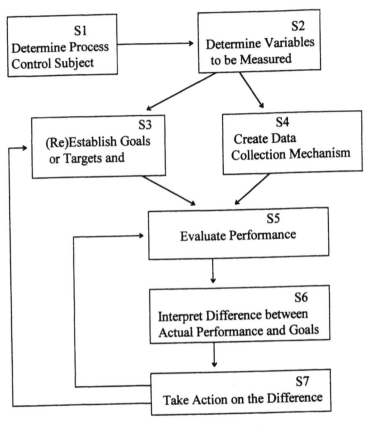

Figure 12.1: Establishing and maintaining quality control.

In the univariate case there is a clear correspondence between the control subject and the monitored variable used for control. Each control subject is being measured thus producing the actual performance data on the monitored variable. In multivariate quality control such a correspondence is not as obvious. If we decide to monitor variables in p dimensions, say, not all measurements must come from the same control subjects. This is particularly relevant when measurements are carried out at several locations. For example some tests might be performed in a remote special laboratory while others are easy to perform on site and produce readily available results. In such circumstances the observation vector can consist of data collected at different times, by different measurement devices and on different control subjects. Choosing the control subjects for multivariate quality control therefore has to take into consideration the needs of the process being monitored as well as the measurement system and logistic capabilities of the process environment.

S2. Determination of the variables to be measured for quality control.

The process capability study described in Chapter 4 produces an understanding of the structure of the data collected on the process being studied. Part of this understanding leads to an identification of key variables to be used in the quality control feedback loop.

S3. Establishment of a goal for the control subject through internally or externally assigned targets for the measured variables.

Chapters 3-5 cover three alternatives to setting targets for the monitored variables. If a previously collected sample is used as a reference sample then quality control determines if the observation being tested is consistent with the process that generated the reference sample. Internally computed targets are used in the context of process capability studies for signalling unusual, outlying observations. Externally assigned targets are derived from considerations outside the process such as standards or competitive benchmarks. Here quality control is designed to establish adherence to such standards or to meet goals set by other competitive processes.

S4. Creation of a sensor or data collection mechanism to evaluate actual performance.

Referring back to the comments on step 1 it is worth emphasizing the complex issues of collecting multivariate data in industrial environments. The data collection mechanism has to include the logistics of getting the control subjects to the measurement devices, scheduling the tests and providing the results back to the operator in charge of controlling the process.

S5. Evaluation of actual performance.

Multivariate quality control involves, in come cases, a complex scheduling of tests so that results can be synchronized to arrive at the right place, at the right time. Evaluation of actual performance is dependent on such a synchronized system. Univariate quality control is much simpler since once a variable has been measured and the results have been analyzed the control operation is possible.

S6. Interpretation of the difference between actual performance and goal.

Once a T^2 chart indicates a condition out of control an investigation using either univariate control or simultaneous univariate and multivariate charts, is triggered. This is so because the T^2 chart does not point out the actual cause of discrepancy between actual performance and goal. Chapter 11 discusses several graphical techniques that can be used in such investigations.

S7. Taking action on the difference using prespecified criteria and responses.

The multivariate case entails a much more elaborate set of criteria and responses than the univariate quality control process. A good understanding of cause and effect relationships is necessary for the quality control response to deviation of actual performance from targets to be well founded and systematic. Results from designed experiments mentioned in the process capability study section of chapter 4 can establish such cause and effect relationships. The possible responses have to be determined in the context of the organization structure of the process being monitored. This should include a description of the level of operator enpowerment and what are the recommended actions to be initiated as a result of out-of-control signals.

Milestones of a Strategic Plan for Implementing Multivariate Quality Control

In order to implement the 7 step multivariate quality control process described above we propose a plan including 6 milestones listed as M1-M6. Specific organizations can use this plan by adding a time table and assign individuals responsible to meet these milestones.

M1. Perform a multivariate process capability study and set targets using a process capability study team.

M2. Design the data collection and feedback system to meet the needs of the process operators and their management.

M3. Develop criteria for action on the differences between actual performance and goal by specifying who is entitled to take what action under what conditions.

M4. Establish a review system to monitor the quality control process using inputs from the process internal customers and process audits.

M5. Pilot the quality control process and provide training to operators and management.

M6. Address resistance to change by planning the transfer to on-going operations of the quality control process.

This plan will enable an organization to implement a multivariate quality control process that operates according to steps 1-7 listed above. A sample plan is presented as Figure 12.2. We conclude with a highlight of main points presented in the text that are specific to multivariate quality control:

1. Hotelling's T^2 is a basic tool for multivariate quality control. With grouped data it can be decomposed into an overall measure of distance of the group means from the target, T_M^2, and an internal measure of variability, T_D^2 (Chapters 1-5).

2. The analysis of T^2-charts needs to be complemented with additional graphical displays (Chapters 6 and 11).

3. The detection of the variables responsible for a multivariate out of control signal can be performed either by graphical univariate methods or by multivariate decomposition of the T_M^2 (Chapters 7 and 11).

4. A large number of variables (dimensions) relative to the number of observations will produce stable T^2-charts (Chapter 4).

5. We distinguish between multivariate quality control with targets from a previously collected references sample (Chapter 5) from

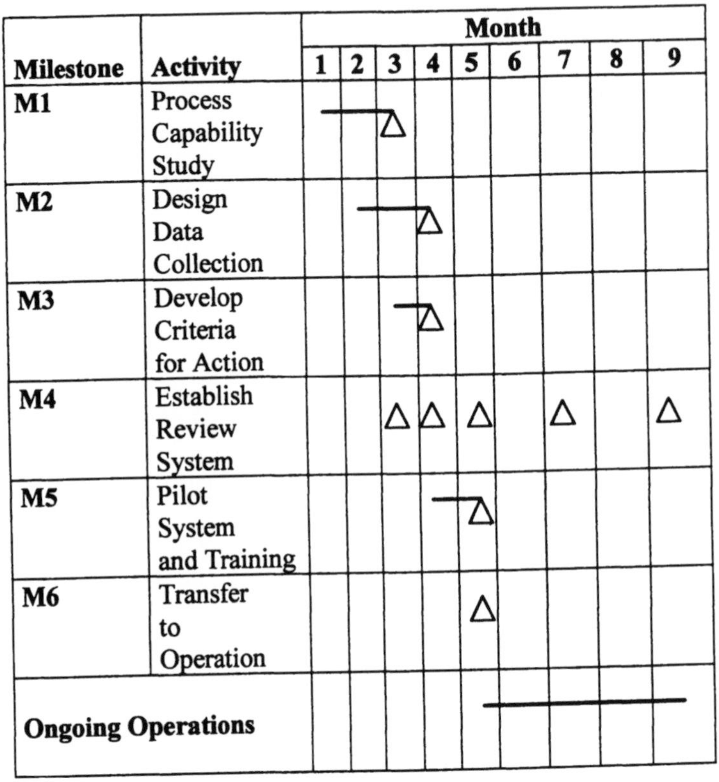

Milestone	Activity	Month								
		1	2	3	4	5	6	7	8	9
M1	Process Capability Study	△								
M2	Design Data Collection			△						
M3	Develop Criteria for Action			△						
M4	Establish Review System				△	△	△	△		△
M5	Pilot System and Training					△				
M6	Transfer to Operation					△				
Ongoing Operations										

Figure 12.2: A sample plan for institutionalizing quality control.

internally assigned targets (Chapter 4) or from externally assigned targets (Chapter 3).

6. Statistical tolerance regions offer an alternative to the Shewhart control chart approach to quality control. Statistical tolerance regions are particularly useful in the multivariate case (Chapter 8).

7. Curtailing inspection of samples from batches can carry many advantages in multivariate quality control. Data collected from batches also allows for a comparison of variability within and between batches with the I_1 and I_2 indices (Chapter 9).

8. Principal components can be useful for multivariate quality control, especially when they lead to new dimensions that can be naturally interpreted (Chapter 10).

9. Graphical techniques available through modern computer technology provide efficient and effective environments for multivariate quality control (Chapter 11).

10. Implementing multivariate quality control requires a plan. Chapter 12 offers some basic elements of such a plan including milestones that have proven essential in successful implementations.

Appendix 1

MINITAB ™ Macros for Multivariate Quality Control

APPENDIX 1.1: POOLED.MTB.

```
note   this is the macro pooled.mtb for grouped data
note   the number of variables in K2;
note   group size in K3;
note   the data are originally in c1-ck2;
note   from the original data set
note   the first 100 observations are copied into
note   c101-ck110.
note   number of observations in data set is in K10
note   number of groups in data set is in K11
       let k2=4
       let k3=2
       let k110=100+k2
       let k130=120+k2
       let k610=600+k2
       let k31=1
       let k32=k31+k3-1
       copy c1-ck2 c101-ck110;
       use 1:100.
```

```
count c101 k10
let k11=k10/k3
set c600
k2(0)
end
let k601=601
store
let ck601=c600
let k601=k601+1
end
exec k2
copy c601-ck610 m2
store
copy c101-ck110 c121-ck130;
use k31:k32.
covariance c121-ck130 m1
print m1
add m1 m2 m2
let k31=k31+k3
let k32=k31+k3-1
end
exec k11
let k12=1/k11
mult m2 k12 m3
print m3
```

APPENDIX 1.2: Macro for Analysis of
Ungrouped Multivariate Data
UU.MTB - The Main Macro

```
note    this is the UU.Mtb program for ungrouped data
note    Tsqbtugr.Mtb performs T-squares on UNGROUPED base and
tested sample
note    data in c1-c9
note    base sample in c41-c59
note    tested sample in c21-c39
note    overall means of base in m50
note    the number of variables in K2
```

```
note
note ENTER HERE NUMBER OF VARIABLES (INTO K2):
let k2=6
note ENTER HERE ROW NUMBER FOR FIRST OBS. IN BASE
note (INTO K971):
let k971=1
note ENTER HERE ROW NUMBER FOR LAST OBS. IN BASE
note (INTO K972):
let k972=30
note ENTER HERE ROW NUMBER FOR FIRST OBS. IN TESTED
note (INTO K973):
let k973=31
note ENTER HERE ROW NUMBER FOR LAST OBS. IN TESTED
note (INTO K974):
let k974=705
let k710=700+k2
note ENTER HERE ZEROS INTO C700 IF THE BASE GRAND
note MEANS ARE
note TO BE COMPUTED FROM SAMPLE
note OTHERWISE - ENTER THE EXTERNAL VALUES:
set c700
k2(0)
end
exec 'inituu.mtb' 1
store
copy c700 ck706;
use k806.
let c717=abs(ck706)
let c718=sum(c717)
let k706=k706+1
let k806=k806+1
end
exec k2
let c719=(c718 gt 0)
copy c719 k719
print k719
copy c1-ck10 c41-ck30;
```

```
use k971:k972.
count c41 k4
set c80
1:k4
end
copy c1-ck10 c21-ck20;
use k973:k974.
count c21 k500
set c60
1:k500
end
let k706=706
note c61-ck40 and m50 has the base mean x-bar-bar
store
mean ck21 ck31
let ck31=(1-k719)*ck31+k719*ck706
std ck21 ck61
let ck71=(ck21-ck31)/ck61
let k21=k21+1
let k31=k31+1
let k61=k61+1
let k71=k71+1
let k706=k706+1
end
exec k2
copy c61-ck40 m50
print c61-ck40
note now tests for base
exec 'tsbsug1.mtb' 1
let c180=c160
erase c160
let c181=((k500-2)*c180)/((k500-1)-(k500/(k500-1))*c180)
note now tests for Tested
let k11=21
let k31=61
let k61=121
let k81=161
```

```
store
let ck81=(ck11-ck31)/ck61
let k11=k11+1
let k31=k31+1
let k61=k61+1
let k81=k81+1
end
exec k2
exec 'tstsug1.mtb' 1
let c183=c160
erase c160
note now performs T-sq with Leave-one-out from c21-ck30
note means and covariance with "left-out" with data in c101-ck110
note the T-square results in c93
let k11=k4
exec 'tsqugloo.mtb' k4
let c182=c160
note: The X-bar-bar, the S - matrix, and the S (-1) - matrix for Base
print c61-ck40
print m18 m15
note: Univariate t-statistics for each variable and observation in Base
print c141-ck80
note: Multivariate TýM , Tý'M and Tý"M (LOO) for each observation
note: in Base
print c180-c182
note: Univariate t-statistics for each variable and observation in Tested
print c161-ck90
note: Multivariate TýM for each observation in Tested
print c60 c183
exec 'uclbtung.mtb' 1
```

INITUU.MTB

```
note    this is the inituu.Mtb program for ungrouped data
let k991=10
let k993=1
store
let kk991=(k993>12)*20+(k993-1)*20+k2
```

```
let k991=k991+10
let k993=k993+1
end
exec 21
let k992=21
let k993=2
store
let kk992=k993*20+1
let k992=k992+10
let k993=k993+1
end
exec 14
let k701=701
let k32=0
let k706=706
let k806=1
```

MSCOVUG.MTB
```
note    Mscovug.Mtb is a routine of Tsbsug1.Mtb, Tstsug1.Mtb
note            and of Tsquglool.Mtb in UU.Mtb
note
count c201 k300
let k101=201
let k111=221
let k121=241
let k110=200+k2
let k120=220+k2
let k130=240+k2
let k140=280+k2
let k141=281
store
mean ck101 ck111
stack ck111 ck141 ck141
let ck121=ck101-ck111
let k101=k101+1
let k111=k111+1
let k121=k121+1
```

```
let k141=k141+1
end
exec k2
copy c221-ck120 m17
copy c241-ck130 m11
trans m11 m12
mult m12 m11 m13
let k299=1/(k300-1)
mult m13 k299 m18
invert m18 m15
```

TSBSUG.MTB
```
note    This is Tsbsug.Mtb, a routine of UU.Mtb
note
copy c41-ck30 c201-ck110
exec 'mscovug.mtb' 1
let k11=k4
copy c41-ck30 c201-ck110
exec 'tsq1.mtb' k4
```

TSQ1.MTB
```
note    Tsq1.Mtb is a routine of Tsbsug1.Mtb in UU.Mtb
note
copy c201-ck110 c421-ck210;
use c80=k11.
let k51=101
let k211=421
let k111=221
store
let ck51=ck211-ck111
let k51=k51+1
let k211=k211+1
let k111=k111+1
end
exec k2
copy c101-ck60 m23
trans m23 m24
```

```
mult m23 m15 m25
mult m25 m24 m26
copy m26 c100
let k11=k11-1
stack c100 c160 c160
```

TSQLOO1.MTB

```
note    Tsqloo1.Mtb is a routine of Tsqugloo.Mtb in UU.Mtb
note
let k51=101
let k211=421
let k111=221
store
let ck51=ck211-ck111
let k51=k51+1
let k211=k211+1
let k111=k111+1
end
exec k2
copy c101-ck60 m23
trans m23 m24
mult m23 m15 m25
mult m25 m24 m26
copy m26 c100
let k11=k11-1
stack c100 c160 c160
```

TSQTS1.MTB

```
note    Tsqts1.Mtb is a routine of Tstsug1.Mtb in UU.Mtb
note
copy c201-ck110 c421-ck210;
use c60=k11.
let k51=101
let k211=421
let k111=221
store
let ck51=ck211-ck111
```

```
let k51=k51+1
let k211=k211+1
let k111=k111+1
end
exec k2
copy c101-ck60 m23
trans m23 m24
mult m23 m15 m25
mult m25 m24 m26
copy m26 c100
let k11=k11-1
stack c100 c160 c160
```

TSQUGLOO.MTB

```
note    Tsqugloo.Mtb is a routine of UU.Mtb
note
copy c41-ck30 c201-ck110;
omit c80=k11.
exec 'mscovug.mtb' 1
mult m15 1 m33
copy c41-ck30 c421-ck210;
use c80=k11.
exec 'tsqloo1.mtb' 1
```

TSTSUG1.MTB

```
note    Tstsug1.Mtb is a routine of UU.Mtb
copy c41-ck30 c201-ck110
exec 'mscovug.mtb' 1
let k11=k500
copy c21-ck20 c201-ck110
exec 'tsqts1.mtb' k500
```

UCLBTUNG.MTB

```
note    Uclbtung.Mtb is a routine of UU.Mtb
note
set c400
.975 .9975 .99865
```

```
end
let k201=(k4-1)
InvCDF C400 c601;
    T K201.
set c400
.95 .995 .9973
end
let k201=(k4-1)*(k4-1)/k4
let k202=k2/2
let k203=(k4-k2-1)/2
InvCDF C400 c404;
    Beta K202 K203.
let c621=c404*k201
let k211=(k4-1)*(k4-2)*k2/(k4*(k4-k2-1))
let k212=k2
let k213=(k4-k2-1)
InvCDF C400 c414;
    F K212 K213.
let c641=c414*k211
let k221=k4*(k4-2)*k2/((k4-1)*(k4-k2-1))
let k222=k2
let k223=(k4-k2-1)
InvCDF C400 c444;
    F K222 K223.
let c661=c444*k221
let k231=(k4+1)*(k4-1)*k2/(k4*(k4-k2))
let k232=k2
let k233=(k4-k2)
InvCDF C400 c464;
    F K232 K233.
let c681=c464*k231
print c400 c601 c621 c641 c661;
format (1x,f9.4,4f10.3).
print c400 c681;
format (1x,f9.4,f10.3).
```

APPENDIX 1.3: MACRO FOR ANALYSIS OF GROUPED MULTIVARIATE DATA

GG.MTB - The Main Macro

```
note   this is the gg.mtb for grouped data
note   the tsqbsgrp macro performs T-squares on GROUPED base and
note   tested sample
note   data in c1-c9 base sample in c21-c29; tested sample in c11-c19;
note   overall means of base in m50
note   the number of variables in K2 ; group size in K3;
note   number of observ in base is in K10; # of observ in tested is in
note   K20;
note   number of groups in base is in K11; # of groups in tested is in
note   K21;
note ENTER HERE NUMBER OF VARIABLES (INTO K2):
let k2=6
note ENTER HERE GROUP SIZE (INTO K3):
let k3=2
note ENTER HERE ROW NUMBER FOR FIRST OBS. IN BASE
note    (INTO K971):
let k971=1
note ENTER HERE ROW NUMBER FOR LAST OBS. IN BASE
note    (INTO K972):
let k972=30
note ENTER HERE ROW NUMBER FOR FIRST OBS. IN TESTED
note    (INTO K973):
let k973=31
note ENTER HERE ROW NUMBER FOR LAST OBS. IN TESTED
note    (INTO K974):
let k974=70
let k9=0+k2
let k110=100+k2
let k210=200+k2
copy c1-ck9 c101-ck110;
use k971:k972.
count c101 k10
copy c1-ck9 c201-ck210;
```

```
use k973:k974.
count c201 k20
exec 'initgg.mtb' 1
store
mean ck101 ck131
let ck151=ck101-ck131
let ck251=ck201-ck131
let k101=k101+1
let k201=k201+1
let k131=k131+1
let k151=k151+1
let k251=k251+1
end
exec k2
note c131-ck140 has the overall mean x-bar-bar and m50 the S(-1)
copy c151-ck160 m11
trans m11 m12
mult m12 m11 m13
let k299=1/(k10-1)
mult m13 k299 m18
invert m18 m50
let k11=k10/k3
let k21=k20/k3
exec 'groupbs.mtb' 1
erase c401-c499
erase c80
note m51 has the sum of (n-1)*S(j) 's and m52 has Sp(-1)
count c201 k20
let k21=k20/k3
set c12
1:k10
end
set c13
1:k11
end
set c22
1:k20
```

```
end
set c23
1:k21
end
exec 'groupts.mtb' 1
erase c401-c499
erase c80
exec 'tsqbsgrp.mtb' 1
let c506=((k11-2)*c505)/(k11-1-c505*k11/(k11-1))
erase c401-c499
erase c80
exec 'tsqtsgrp.mtb' 1
erase c401-c499
erase c80
note now performs T-sq with Leave-one-out from c101-ck110
note means and covariance with "left-out" with data in c101-ck110
note the T-square results in c507
let k406=k10
let k405=k10-k3+1
let k407=k11
exec 'tsqgrloo.mtb' k11
let c507=k3*c80
exec 'uclsbtgr.mtb' 1
note c200 alphas-c301 c311 c321 c331 UCL's with S - matrix
note (beta+two F's for Base+one for Tested)
print c200 c301 c311 c321 c331;
format (1x,f9.4,4f10.3).
exec 'uclspbtg.mtb' 1
note c200 alphas- Ucls's - c311 for Base, 321 for tested with Sp - matrix
print c200 c311 c321;
format (1x,f9.4,2f10.3).
let k506=k10
let k505=k506-k3+1
store
copy c501 c502 c511 c512 c901 c902 c911 c912;
use k505:k506.
sum c901 k903
```

sum c902 k904
sum c911 k913
sum c912 k914
stack k903 c503 c503
stack k904 c504 c504
stack k913 c513 c513
stack k914 c514 c514
let k505=k505-k3
let k506=k506-k3
end
exec k11
let k506=k20
let k505=k506-k3+1
store
copy c601 c602 c611 c612 c901 c902 c911 c912;
use k505:k506.
sum c901 k903
sum c902 k904
sum c911 k913
sum c912 k914
stack k903 c603 c603
stack k904 c604 c604
stack k913 c613 c613
stack k914 c614 c614
let k505=k505-k3
let k506=k506-k3
end
exec k21
note: X-bar-bar, the S-matrix, S(-1)-matrix, sum of(n-1)S(j) & Sp(-1)
for Base
print c131-ck140
print m18 m50 m51 m52
note For Base with S:$T^2$0(groups)in c503 T^2D(groups)in c504 T^2M
(groups)in c505
note T^2M′ (groups) in c506 T^2M″ (groups - LOO) in c507
print c503-c507
note For Base with Sp:$T^2$0(groups)in c513 T^2D(groups)in c514 T^2M

note (groups)in c515
print c513-c515
note For Tested with S:T^{2}0(groups)in c603 T^{2}D(groups)in c604 T^{2}M
note (groups)in c605
print c603-c605
let c617=c614/k3
note ForTested with Sp:T^{2}0(groups)in c613 T^{2}D(groups)in c614 T^{2}M
note (groups)in c615
note ForTested with Sp:T^{2}D(groups) divided by obs. per group in c617
print c613-c615 c617
note c200 alphas-c301 c311 c321 c331 UCL's with S - matrix
note (beta+two F's for Base+one for Tested)
print c200 c301 c311 c321 c331;
format (1x,f9.4,4f10.3).
InvCDF c200 c327;
ChiSquare k3.
note c200 alphas c327 are the Chi-square values as UCL for T^{2}D
print c200 c327;
format (1x,f9.4,f10.3).

INITGG.MTB
note this is the Initgg.mtb routine for grouped data
let k991=120
let k993=1
store
let kk991=k991-10+k2
let k991=k991+10
let k993=k993+1
end
exec 18
let k410=400+k2
let k992=101
let k993=1
store
let kk992=k992
let k992=k992+10
let k993=k993+1

end
exec 20

GMEANSTS.MTB
copy c701-ck710 c401-ck410;
use k405:k406.
let k407=k406-k405+1
let k100=110
store
let k999=k100+10
let k998=k100-9
let kk999=k999-10+k2
let kk998=k998
let k100=k100+10
end
exec 88
let k521=521
let k531=531
let k541=541
let k551=551
let k561=561
let k530=521+k2
let k540=531+k2
let k550=541+k2
let k560=551+k2
let k570=561+k2
store
mean ck401 ck411
std ck401 c528
stack ck411 ck441 ck441
stack c528 ck521 ck521
let ck421=ck401-ck411
let ck461=ck411-ck131
let ck551=sqrt(k407)*(ck411-ck131)/ck541
stack ck421 ck451 ck451
stack ck461 ck471 ck471
stack ck551 ck561 ck561

```
let k401=k401+1
let k421=k421+1
let k131=k131+1
let k441=k441+1
let k451=k451+1
let k461=k461+1
let k471=k471+1
let k521=k521+1
let k531=k531+1
let k541=k541+1
let k551=k551+1
let k561=k561+1
end
exec k2
let k405=k405-k3
let k406=k406-k3
let k521=521
let k531=531
let k541=541
let k551=551
let k561=561
let k571=571
let k581=581
let k591=591
```

GROUPBS.MTB
```
let k403=k11
let k406=k10
let k405=k406-k3+1
let k401=401
let k411=411
let k421=421
let k131=131
let k441=441
let k451=451
let k461=461
let k541=541
```

```
let k410=400+k2
let k420=410+k2
let k430=420+k2
let k440=430+k2
let k450=440+k2
let k460=450+k2
let k470=460+k2
let k480=470+k2
let k710=700+k2
let k610=600+k2
let k550=540+k2
set c600
k2(0)
end
let k601=601
store
let ck601=c600
let k601=k601+1
end
exec k2
copy c601-ck610 m14
erase c601-ck610
copy c101-ck110 c701-ck710
exec 'mscov.mtb' k403
invert m18 m15
copy c441-ck450 c111-ck120
copy c451-ck460 c161-ck170
copy c471-ck480 c171-ck180
copy m14 m51
diag m18 c558
let c559=sqrt(c558)
copy c559 m60
trans m60 m61
copy m61 c541-ck550
copy m15 m52
copy m13 m53
```

GROUPTS.MTB
```
let k403=k21
let k406=k20
let k405=k406-k3+1
let k100=110
store
let k999=k100+10
let k998=k100-9
let kk999=k999-10+k2
let kk998=k998
let k100=k100+10
end
exec 88
copy c201-ck210 c701-ck710
exec 'gmeansts.mtb' k403
let k521=521
let k531=531
let k571=571
let k581=581
let k591=591
store
mean ck531 ck581
std ck531 ck591
let ck571=(ck521-ck581)/ck591
let k521=k521+1
let k531=k531+1
let k571=k571+1
let k581=k581+1
let k591=k591+1
end
exec k2
copy c441-ck450 c211-ck220
copy c451-ck460 c261-ck270
copy c471-ck480 c271-ck280
let k531=531
let k571=571
```

```
let k581=581
let k591=591
```

MSCOV.MTB
```
copy c701-ck710 c401-ck410;
use k405:k406.
count c401 k400
let k401=401
let k411=411
let k421=421
let k131=131
let k441=441
let k451=451
let k461=461
let k471=471
let k531=531
store
mean ck401 ck411
std ck401 c529
stack ck411 ck441 ck441
stack c529 ck531 ck531
let ck421=ck401-ck411
let ck461=ck411-ck131
stack ck421 ck451 ck451
stack ck461 ck471 ck471
let k401=k401+1
let k411=k411+1
let k421=k421+1
let k131=k131+1
let k441=k441+1
let k451=k451+1
let k461=k461+1
let k471=k471+1
let k531=k531+1
end
exec k2
let k531=531
```

```
copy c421-ck430 m11
trans m11 m12
mult m12 m11 m13
add m13 m14 m14
let k299=1/(k403*(k400-1))
mult m14 k299 m18
let k405=k405-k3
let k406=k406-k3
```

TSQBSGRP.MTB
```
erase c401-c499
 erase c80
 copy c151-ck160 c401-ck410
 copy m50 m60
 let k600=k10
 let c70=c12
 exec 'tsqnew.mtb' k10
 let c501=c80
 erase c401-c499
 erase c80
 let k600=k10
 let c70=c12
 copy c161-ck170 c401-ck410
 copy m50 m60
 let k600=k10
 let c70=c12
 exec 'tsqnew.mtb' k10
 let c502=c80
 erase c401-c499
 erase c80
 copy c171-ck180 c401-ck410
 copy m50 m60
 let k600=k11
 let c70=c13
 exec 'tsqnew.mtb' k11
 let c505=k3*c80
 erase c401-c499
```

```
erase c80
copy c151-ck160 c401-ck410
copy m52 m60
let k600=k10
let c70=c12
exec 'tsqnew.mtb' k10
let c511=c80
erase c401-c499
erase c80
copy c161-ck170 c401-ck410
copy m52 m60
let k600=k10
let c70=c12
exec 'tsqnew.mtb' k10
let c512=c80
erase c401-c499
erase c80
copy c171-ck180 c401-ck410
copy m52 m60
let k600=k11
let c70=c13
exec 'tsqnew.mtb' k11
let c515=k3*c80
erase c401-c499
erase c80
```

TSQBSNEW.MTB
```
copy c411-ck420 m23
trans m23 m24
mult m23 m60 m25
mult m25 m24 m26
copy m26 c50
```

TSQGRLOO.MTB
```
let k801=801
let k811=811
let k821=821
```

```
let k831=831
let k841=841
let k851=851
let k810=800+k2
let k820=810+k2
let k420=410+k2
let k830=820+k2
let k840=830+k2
let k850=840+k2
let k860=850+k2
copy c101-ck110 c801-ck810;
use c12=k405:k406.
count c801 k400
store
mean ck801 ck811
let k801=k801+1
let k811=k811+1
end
exec k2
copy c101-ck110 c821-ck830;
omit c12=k405:k406.
let k801=801
let k811=811
let k821=821
let k831=831
let k841=841
let k851=851
store
mean ck821 ck831
let ck841=ck821-ck831
let ck851=ck811-ck831
let k811=k811+1
let k821=k821+1
let k831=k831+1
let k841=k841+1
let k851=k851+1
end
```

```
exec k2
copy c841-ck850 m11
count c841 k400
trans m11 m12
mult m12 m11 m13
let k299=1/(k400-1)
mult m13 k299 m18
invert m18 m15
  copy c851-ck860 c411-ck420
  copy m15 m60
  exec 'tsqbsnew.mtb' 1
  stack c50 c80 c80
  let k405=k405-k3
  let k406=k406-k3
  let k407=k407-1
```

TSQNEW.MTB
```
copy c401-ck410 c411-ck420;
use c70=k600.
copy c411-ck420 m23
trans m23 m24
mult m23 m60 m25
mult m25 m24 m26
copy m26 c50
let k600=k600-1
stack c50 c80 c80
```

UCLSBTGR.MTB
```
set c200
.95 .995 .9973
end
let k701=(k11-1)*(k11*k3-1)/(k11)
let k702=k2/2
let k703=(k11*k3-k2-1)/2
InvCDF C200 c704;
    Beta K702 K703.
```

let c301=c704*k701
let k711=(k11-1)*(k11*k3-2)*k2/(k11*(k11*k3-k2-1))
let k712=k2
let k713=(k11*k3-k2-1)
InvCDF C200 c714;
    F K712 K713.
let c311=c714*k711
let k721=k11*(k11*k3-2)*k2/((k11-1)*(k11*k3-k2-1))
let k722=k2
let k723=(k11*k3-k2-1)
InvCDF C200 c724;
    F K722 K723.
let c321=c724*k721
let k731=(k11+1)*(k11*k3-1)*k2/(k11*(k11*k3-k2))
let k732=k2
let k733=(k11*k3-k2)
InvCDF C200 c734;
    F K732 K733.
let c331=c734*k731

UCLSPBTG.MTB
set c200
.95 .995 .9973
end
let k711=(k11-1)*k11*(k3-1)*k2/(k11*(k11*(k3-1)-k2+1))
let k712=k2
let k713=(k11*(k3-1)-k2+1)
InvCDF C200 c714;
    F K712 K713.
let c311=c714*k711
let k721=(k11+1)*k11*(k3-1)*k2/(k11*(k11*(k3-1)-k2+1))
let k722=k2
let k723=(k11*(k3-1)-k2+1)
InvCDF C200 c724;
    F K722 K723.
let c321=c724*k721

APPENDIX 1.4: MACRO FOR CREATING MP-CHART FOR UNGROUPED DATA
MPGRAPH.MTB — THE MAIN MACRO

```
note This is the main MpGraph program
note Base calculations are in file UU.Mtb
note
erase c245-c250
erase c801-c806
let c801=(c81>4)*4+(c81<-4)*(-4)+(c81 ge -4 and c81 le 4)*c81
let c802=(c82>4)*4+(c82<-4)*(-4)+(c82 ge -4 and c82 le 4)*c82
let c803=(c83>4)*4+(c83<-4)*(-4)+(c83 ge -4 and c83 le 4)*c83
let c804=(c84>4)*4+(c84<-4)*(-4)+(c84 ge -4 and c84 le 4)*c84
let c805=(c85>4)*4+(c85<-4)*(-4)+(c85 ge -4 and c85 le 4)*c85
let c806=(c86>4)*4+(c86<-4)*(-4)+(c86 ge -4 and c86 le 4)*c86
let c903=(c93>c341(3))*(c341(3)+4)+(c93 le c341(3))*c93
count c801 k20
exec 'mpdat1.mtb' k20
let c248=c247+c245
let c250=(abs(c247)>3)*1+(abs(c247) le 3 and
     abs(c247)>2)*4+(abs(c247) le 2 and abs(c247)>1)*13
let k244=c341(1)
let k245=c341(2)
let k246=c341(3)
chart sum(c248)*c246;
  cluster c249;
  bar;
    base c245;
    type 1
    color c250;
  tick 2 k244 k245 k246;
    TSize 1;
    TFont 1;
    label 'T²90' 'T²95' 'T²99973';
  tick 1;
    TSize 1;
  GRID 2;
    title 'MP - Chart for Variables: Case Study 1';
```

```
   TFont 1;
 axis 2;
   TFont 1;
   label 'Univariate and Multivariate Statistics';
 axis 1;
   TFont 1;
   label 'Groups'.
end
```

MPDAT1.MTB

```
note MpDat1 is a routine of MpGraph.Mtb
note
let k11 = 6
exec 'mpdat.mtb' k11
let k20=k20-1
end
```

MPDAT.MTB

```
note MpDat is a loop routine of MpDat1.Mtb in MpGraph.Mtb
note
let k13=k11+800
copy ck13 c7;
use k20.
stack c7 c247 c247
copy c903 c7;
use k20.
stack c7 c245 c245
copy k11 c7
stack c7 c249 c249
copy k20 c7
stack c7 c246 c246
let k11=k11-1
```

**APPENDIX 1.5: MACRO FOR CREATING MP-CHART
FOR GROUPED DATA
MPGROUP.MTB—THE MAIN MACRO FOR MP CHART FOR
MEANS**

note This is the main MpGraph program
note It displays the MpGraph for 6 variables
note Data is in file Case1sm6.Mtw in the Data folder
note Base calculations are in file GG.Mtb
note (Calculated data is in file Case1sm6_Tmp.Mtw)
note
note ENTER HERE NUMBER OF VARIABLES (INTO K2):
let k2=6
note ENTER HERE NUMBER OF FIRST COLUMN IN THE SERIES
note OF K2 COLUMNS OF DATA (INTO K973):
let k973=561
note ENTER HERE NUMBER OF COLUMN CONTAINING THE T2
note VALUES (INTO K974):
let k974=605
note ENTER HERE NUMBER OF COLUMN CONTAINING THE 3
note CRITICAL VALUES IN ASCENDING ORDER (INTO K975):
LET K975=321
note ENTER HERE NUMBER OF FIRST EMPTY COLUMN IN A
note SERIES OF K2 EMPTY COLUMNS (INTO K972):
let k972=801
note ENTER HERE NUMBER OF FIRST EMPTY COLUMN IN A
note SERIES OF 6 EMPTY COLUMNS (INTO K991):
let k991=245
note ENTER HERE NUMBER OF FIRST EMPTY COLUMN IN A
note SERIES OF 2 EMPTY COLUMNS (INTO K976):
let k976=903
note
let k977=k976+1
let k992=k991+1
let k993=k991+2
let k994=k991+3
let k995=k991+4
let k996=k991+5

```
let k982=k972+k2-1
let k983=k973+k2-1
erase ck991-ck996
erase ck972-ck982
store
let ck972=(ck973>4)*4+(ck973<-4)*(-4)+(ck973 ge -4 and ck973 le
4)*ck973
let k972=k972+1
let k973=k973+1
end
exec k2
let k972=k972-k2
let k973=k973-k2
let ck976=(ck974>ck975(3))*(ck975(3)+4)+(ck974 le ck975(3))*ck974
count ck972 k20
exec 'mpdat1.mtb' k20
let ck994=ck993+ck991
let ck996=(abs(ck993)>3)*1+(abs(ck993) le 3 and
     abs(ck993)>2)*4+(abs(ck993) le 2 and abs(ck993)>1)*13
let k244=ck975(1)
let k245=ck975(2)
let k246=ck975(3)
chart sum(ck994)*ck992;
 cluster ck995;
 bar;
  base ck991;
  type 1;
  color ck996;
 tick 2 k244 k245 k246;
  TSize 1;
  TFont 1;
  label 'T²95' 'T²995' 'T²9973';
 tick 1;
  TSize 1;
 GRID 2;
 axis 2;
  TFont 1;
```

    label 'Univariate and Multivariate Statistics ';
  axis 1;
    TFont 1;
    label 'Groups'.
end

MPGDISP.MTB—THE MAIN MACRO FOR MP CHART FOR DISPERSION

note This is the main MpGrDisp program
note It displays the MpGraph for Dispersion k2 variables
note Base calculations are in file GG.Mtb
note
note ENTER HERE NUMBER OF VARIABLES (INTO K2):
let k2=6
note ENTER HERE NUMBER OF FIRST COLUMN IN THE SERIES
note OF K2 COLUMNS OF DATA (INTO K973):
let k973=571
note ENTER HERE NUMBER OF COLUMN CONTAINING
note THE (T2D/K3) VALUES (INTO K974):
let k974=617
note ENTER HERE NUMBER OF COLUMN CONTAINING
note THE 3 CRITICAL VALUES IN ASCENDING ORDER
note (INTO K975):
LET K975=327
note ENTER HERE NUMBER OF FIRST EMPTY COLUMN IN
note A SERIES
note OF K2 EMPTY COLUMNS (INTO K972):
let k972=801
note ENTER HERE NUMBER OF FIRST EMPTY COLUMN IN
note A SERIES
note OF 6 EMPTY COLUMNS (INTO K991):
let k991=245
note ENTER HERE NUMBER OF FIRST EMPTY COLUMN IN
note A SERIES
note OF 2 EMPTY COLUMNS (INTO K976):
let k976=903

```
note
let k977=k976+1
let k992=k991+1
let k993=k991+2
let k994=k991+3
let k995=k991+4
let k996=k991+5
let k982=k972+k2-1
let k983=k973+k2-1
erase ck991-ck996
erase ck972-ck982
store
let ck972=(ck973>4)*4+(ck973<-4)*(-4)+(ck973 ge -4 and ck973
     le 4)*ck973
let k972=k972+1
let k973=k973+1
end
exec k2
let k972=k972-k2
let k973=k973-k2
let ck976=(ck974>ck975(3))*(ck975(3)+4)+(ck974 le ck975(3))*ck974
count ck972 k20
exec 'mpdat1.mtb' k20
let ck994=ck993+ck991
let ck996=(abs(ck993)>3)*1+(abs(ck993) le 3 and
     abs(ck993)>2)*4+(abs(ck993) le 2 and
abs(ck993)>1)*13
let k244=ck975(1)
let k245=ck975(2)
let k246=ck975(3)
chart sum(ck994)*ck992;
 cluster ck995;
 bar;
  base ck991;
  type 1;
  color ck996;
```

```
  tick 2 k244 k245 k246;
    TSize 1;
    TFont 1;
    label 'Chi²95' 'Chi²995' 'Chi²9973';
  tick 1;
    TSize 1;
  GRID 2;
note title 'FIGURE 11.4b';
note TFont 1;
note title 'Case Study 1';
note TFont 1;
  axis 2;
    TFont 1;
    label 'Univariate and Multivariate Statistics ';
  axis 1;
    TFont 1;
    label 'Groups'.
end
```

MPDAT1G.MTB
```
note MpDat1 is a routine of MpGraph.Mtb
note
let k11 = k2
exec 'mpdat.mtb' k2
let k20=k20-1
end
```

MPDATG.MTB
```
note MpDat is a loop routine of MpDat1.Mtw in MpGraph.Mtw
note
let k13=k11+k972-1
copy ck13 ck977;
use k20.
stack ck977 ck993 ck993
copy ck976 ck977;
use k20.
stack ck977 ck991 ck991
```

```
copy k11 ck977
stack ck977 ck995 ck995
copy k20 ck977
stack ck977 ck992 ck992
let k11=k11-1
```

Appendix 2

The Data from
the Case Studies

CASE STUDY 1.

The data in Case Study 1 originate from a process capability study for turning aluminium pins. Six diameter and length measurements were recorded for each of the 70 observations. The first three variables are diameter measurements on three different locations on the main part of the pin, the forth is a diameter measurement at the cap, and the last two are the lengths measurements, without and with the cap, respectively. In the analyses, the first 30 out of the 70 observations were selected as a base sample against which we compare all the observations.

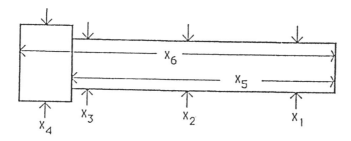

Figure A1: Six critical dimensions of aluminum pins.

TABLE A2.1
The data from Case Study 1

Number	Diameter1	Diameter2	Diameter3	Diameter4	Length1	Length2
1	9.99	9.97	9.96	14.97	49.89	60.02
2	9.96	9.96	9.95	14.94	49.84	60.02
3	9.97	9.96	9.95	14.95	49.85	60.00
4	10.00	9.99	9.99	14.99	49.89	60.06
5	10.00	9.99	9.99	14.99	49.91	60.09
6	9.99	9.99	9.98	14.99	49.91	60.08
7	10.00	9.99	9.99	14.98	49.91	60.08
8	10.00	9.99	9.99	14.99	49.89	60.09
9	9.96	9.95	9.95	14.95	50.00	60.15
10	9.99	9.98	9.98	14.99	49.86	60.06
11	10.00	9.99	9.98	14.99	49.94	60.08
12	10.00	9.99	9.99	14.99	49.92	60.05
13	9.97	9.96	9.96	14.96	49.90	60.02
14	9.97	9.96	9.96	14.96	49.91	60.02
15	9.97	9.97	9.96	14.97	49.90	60.01
16	9.97	9.97	9.96	14.97	49.89	60.04
17	9.98	9.97	9.96	14.96	50.01	60.13
18	9.98	9.97	9.97	14.96	49.93	60.06
19	9.98	9.98	9.97	14.98	49.93	60.02
20	9.98	9.97	9.97	14.97	49.94	60.06
21	9.98	9.97	9.97	14.97	49.93	60.06
22	9.98	9.97	9.97	14.97	49.91	60.02
23	9.98	9.97	9.96	14.98	49.92	60.06
24	10.00	9.99	9.98	14.98	49.88	60.00
25	9.99	9.99	9.99	14.98	49.91	60.04
26	10.00	9.99	9.99	14.99	49.85	60.01
27	10.00	10.00	9.99	14.99	49.91	60.05
28	10.00	9.99	9.99	15.00	49.92	60.04
29	10.00	9.99	9.99	14.99	49.89	60.01
30	10.00	10.00	9.99	14.99	49.88	60.00
31	10.00	9.99	9.99	14.99	49.92	60.03
32	10.00	9.99	9.99	15.00	49.93	60.03
33	10.00	10.00	9.99	14.99	49.91	60.02
34	10.00	9.99	9.99	14.99	49.92	60.02
35	10.00	9.99	9.99	14.99	49.92	60.00
36	10.00	10.00	9.99	15.00	49.94	60.05
37	10.00	9.99	9.99	15.00	49.89	59.98
38	10.00	10.00	9.99	14.99	49.93	60.01

TABLE A2.1 Continued.

Number	Diameter1	Diameter2	Diameter3	Diameter4	Length1	Length2
39	10.00	10.00	9.99	14.99	49.94	60.02
40	10.00	10.00	9.99	15.00	49.86	59.96
41	10.00	9.99	9.99	14.99	49.90	59.97
42	10.00	10.00	10.00	14.99	49.92	60.00
43	10.00	10.00	9.99	14.98	49.91	60.00
44	10.00	10.00	10.00	15.00	49.93	59.98
45	10.00	9.99	9.98	14.98	49.90	59.99
46	9.99	9.99	9.99	14.99	49.88	59.98
47	10.01	10.01	10.01	15.01	49.87	59.97
48	10.00	10.00	9.99	14.99	49.81	59.91
49	10.01	10.00	10.00	15.01	50.07	60.13
50	10.01	10.00	10.00	15.00	49.93	60.00
51	10.00	10.00	10.00	14.99	49.90	59.96
52	10.01	10.01	10.01	15.00	49.85	59.93
53	10.00	9.99	9.99	15.00	49.83	59.98
54	10.01	10.01	10.00	14.99	49.90	59.98
55	10.01	10.01	10.00	15.00	49.87	59.96
56	10.00	9.99	9.99	15.00	49.87	60.02
57	9.99	9.99	9.99	14.98	49.92	60.03
58	9.99	9.98	9.98	14.99	49.93	60.03
59	9.99	9.99	9.98	14.99	49.89	60.01
60	10.00	10.00	9.99	14.99	49.89	60.01
61	9.99	9.99	9.99	15.00	50.04	60.15
62	10.00	10.00	10.00	14.99	49.84	60.03
63	10.00	10.00	9.99	14.99	49.89	60.01
64	10.00	9.99	9.99	15.00	49.88	60.01
65	10.00	10.00	9.99	14.99	49.90	60.04
66	9.90	9.89	9.91	14.88	49.99	60.14
67	10.00	9.99	9.99	15.00	49.91	60.04
68	9.99	9.99	9.99	14.98	49.92	60.04
69	10.01	10.01	10.00	15.00	49.88	60.00
70	10.00	9.99	9.99	14.99	49.95	60.01

CASE STUDY 2.

The data in Case Study 2 originate from another process capability study - of a quite complex mechanical part on which 11 diameter, 4 length and 2 wall width measurements were recorded. The first four variables are inner diameter measurements at four locations of a cylindrical segment of the part, and the next four are the outer diameter measurements at the same locations. Variables 9 and 10 are again two diameter measurements with equal reference value (on another cylindrical segment of the part), while the 11th variable is a diameter measurement at a different location. The four length measurements are of various portions of the part, with the last among them being the overall length. The external reference values for the two wall width measurements (variables 16 and 17) had only an upper critical value (of 0.20).

In various analyses we used 13 variables (by deleting variables 3,4 and 7,8) or 11 variables (by deleting the last two width variables in addition to variables 3,4 and 7,8).

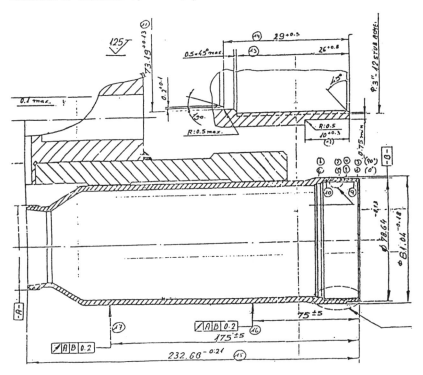

Figure A2: 17 variables of a complex cylindrical part.

TABLE A2.2
The data from Case Study 2

number	var01	var02	var03	var04	var05	var06
1	78.61	78.60	78.55	78.54	80.95	80.93
2	78.52	78.50	78.52	78.51	80.94	80.93
3	78.58	78.55	78.57	78.54	80.90	80.89
4	78.62	78.58	78.55	78.51	80.92	80.92
5	78.61	78.58	78.55	78.52	80.92	80.92
6	78.58	78.55	78.53	78.51	80.90	80.90
7	78.51	78.59	78.51	78.49	80.92	80.93
8	78.54	78.56	78.54	78.53	80.91	80.91
9	78.56	78.55	78.51	78.51	80.91	80.91
10	78.63	78.61	78.59	78.57	80.97	80.95
11	78.62	78.60	78.59	78.57	80.94	80.93
12	78.62	78.60	78.59	78.57	80.95	80.94
13	78.60	78.59	78.58	78.58	80.92	80.92
14	78.64	78.61	78.56	78.53	80.94	80.94
15	78.60	78.58	78.60	78.57	80.93	80.93
16	78.65	78.62	78.60	78.57	80.96	80.96
17	78.58	78.57	78.55	78.54	80.91	80.91
18	78.58	78.56	78.56	78.59	80.91	80.91
19	78.60	78.58	78.58	78.57	80.93	80.92
20	78.61	78.60	78.61	78.59	80.94	80.94
21	78.60	78.59	78.60	78.58	80.93	80.93
22	78.63	78.60	78.61	78.58	80.96	80.96
23	78.60	78.58	78.60	78.58	80.94	80.93
24	78.60	78.58	78.59	78.60	80.94	80.93
25	78.64	78.62	78.61	78.58	80.95	80.95
26	78.60	78.58	78.60	78.58	80.94	80.94
27	78.62	78.59	78.60	78.57	80.95	80.94
28	78.63	78.61	78.58	78.57	80.96	80.94
29	78.63	78.61	78.59	78.57	80.95	80.95
30	78.70	78.68	78.53	78.50	80.99	80.98
31	78.63	78.60	78.61	78.59	80.95	80.94
32	78.64	78.62	78.60	78.58	80.96	80.96
33	78.60	78.58	78.57	78.55	80.93	80.93
34	78.61	78.58	78.56	78.54	80.93	80.92
35	78.62	78.59	78.60	78.58	80.93	80.93
36	78.62	78.60	78.56	78.54	80.94	80.93

Continues

TABLE A2.2 Continued.

number	var07	var08	var09	var10	var11	var12
1	80.90	80.88	74.95	74.95	73.28	10.00
2	80.87	80.87	74.95	75.00	73.27	10.00
3	80.89	80.89	74.93	75.00	73.28	10.00
4	80.85	80.86	74.93	75.00	73.28	10.00
5	80.87	80.87	74.95	75.00	73.27	10.00
6	80.86	80.86	74.95	75.00	73.27	10.00
7	80.85	80.85	74.94	75.00	73.27	10.00
8	80.88	80.88	74.02	75.00	73.27	10.00
9	80.91	80.90	74.95	75.00	73.27	10.00
10	80.92	80.92	74.94	75.00	73.28	10.04
11	80.93	80.93	74.95	74.97	73.29	10.06
12	80.92	80.92	74.95	74.75	73.29	10.04
13	80.92	80.92	74.96	74.75	73.28	10.04
14	80.86	80.85	74.94	74.99	73.28	10.06
15	80.92	80.92	74.93	74.98	73.28	10.06
16	80.92	80.90	74.92	74.99	73.28	10.08
17	80.89	80.90	74.93	74.98	73.25	10.06
18	80.87	80.86	74.92	74.96	73.28	10.06
19	80.91	80.91	74.93	74.99	73.27	10.08
20	80.92	80.92	74.92	74.99	73.28	10.10
21	80.92	80.93	74.93	74.98	73.26	10.10
22	80.93	80.93	74.93	74.99	73.28	10.04
23	80.92	80.92	74.92	74.97	73.27	10.06
24	80.92	80.92	74.92	74.97	73.26	10.04
25	80.93	80.93	74.93	74.99	73.27	10.06
26	80.93	80.93	74.92	74.98	73.27	10.06
27	80.89	80.89	74.91	74.98	73.27	10.04
28	80.92	80.92	74.92	74.99	73.27	10.06
29	80.92	80.91	74.93	74.98	73.28	10.12
30	80.85	80.86	74.94	75.00	73.28	10.08
31	80.94	80.93	74.93	74.99	73.28	10.06
32	80.93	80.93	74.94	75.00	73.27	10.06
33	80.89	80.89	74.93	75.00	73.27	10.04
34	80.90	80.89	74.93	74.99	73.27	10.00
35	80.93	80.93	74.93	74.99	73.26	10.06
36	80.93	80.93	74.94	74.99	73.28	10.04

TABLE A2.2 Continued.

number	var13	var14	var15	var16	var17
1	26.22	29.00	232.60	0.07	0.12
2	26.20	29.00	232.57	0.08	0.13
3	26.18	29.00	232.65	0.05	0.12
4	26.18	29.00	232.54	0.08	0.15
5	26.20	29.00	232.63	0.09	0.14
6	26.20	29.00	232.56	0.10	0.18
7	26.20	29.00	232.57	0.13	0.25
8	26.30	29.00	232.61	0.05	0.15
9	26.22	29.00	232.56	0.07	0.16
10	26.22	29.00	232.61	0.05	0.12
11	26.30	29.00	232.59	0.05	0.13
12	26.20	29.00	232.63	0.06	0.05
13	26.20	29.00	232.56	0.10	0.20
14	26.20	29.00	232.61	0.12	0.16
15	26.20	29.00	232.60	0.12	0.18
16	26.20	29.00	232.55	0.05	0.25
17	26.18	29.00	232.67	0.05	0.05
18	26.24	29.04	232.63	0.06	0.04
19	26.20	29.04	232.58	0.03	0.12
20	26.10	29.08	232.57	0.12	0.17
21	26.16	29.06	232.66	0.08	0.06
22	26.16	29.06	232.50	0.08	0.10
23	26.16	29.06	232.62	0.10	0.05
24	26.22	29.08	232.67	0.04	0.09
25	26.14	28.98	232.60	0.15	0.15
26	26.16	28.98	232.63	0.10	0.05
27	26.20	29.02	232.63	0.04	0.10
28	26.21	29.04	232.67	0.07	0.07
29	26.28	28.98	232.70	0.11	0.08
30	26.22	28.98	232.65	0.05	0.17
31	26.20	29.04	232.66	0.09	0.10
32	26.22	29.04	232.66	0.08	0.13
33	26.22	29.04	232.57	0.04	0.14
34	26.20	28.98	232.61	0.07	0.14
35	26.28	29.04	232.55	0.13	0.30
36	26.25	28.98	232.53	0.13	0.30

CASE STUDY 3.

Raw materials used in the manufacturing of hybrid microcircuits consist of components, dyes, pastes and ceramic substrates. The substrates plates undergo a process of printing and firing during which layers of conductors, dielectric, resistors and platinum or gold are added to the plates. Subsequent production steps consist of laser trimming, mounting and reflow soldering or chip bonding. The last manufacturing stage is the packaging and sealing of complete modules.

Five dimensions of substrate plates are considered here, with labels (a, b, c), (W, L). The first tree are determined by the laser inscribing process. The last two are outer physical dimensions. These outer dimension are measured on a different instrument than the first three, using a different sample of substrates. The two sets of measurements have in common the batch affiliation.

The data in the Case Study 3 are the values of the (a, b, c) dimensions of four lots of ceramic substrates. The first production lot of 13 units (labelled Reference) proved to be of extremely good quality yielding an overall smooth production with no scrap and repairs. This first lot was therefore considered a "standard" to be met by all following lots. Any evidence of lots not meeting this standard should give a signal for action.

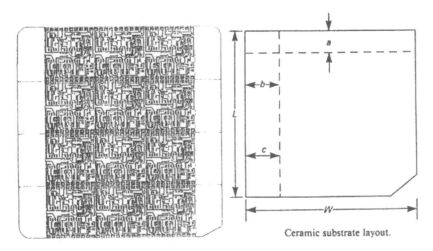

Ceramic substrate layout.

Figure A3: Ceramic substrates used in the microelectronics industry with 5 critical dimensions.

TABLE A2.3

The data from Case Study 3

a	b	c	Sample	W	L
200	552	550	REFERENCE	4001	4002
199	550	550	REFERENCE	4001	4000
197	551	551	REFERENCE	4003	4000
198	552	551	REFERENCE	3999	3998
199	552	551	REFERENCE	3997	4000
199	550	551	REFERENCE	4002	4001
200	550	551	REFERENCE	3998	4000
200	551	552	REFERENCE	3999	4000
200	550	552	REFERENCE	3998	4000
200	550	551	REFERENCE	4000	4002
198	550	551	REFERENCE	4001	4000
198	550	550	REFERENCE	4000	3997
199	550	551	REFERENCE	4001	4000
200	550	550	SAMPLE 30	4002	3998
201	501	550	SAMPLE 30	4000	4001
201	550	550	SAMPLE 30	4001	3998
201	550	501	SAMPLE 30	4001	3999
200	551	552	SAMPLE 7	4005	4000
192	551	552	SAMPLE 7	4010	3985
192	553	552	SAMPLE 7	4005	3995
199	553	550	SAMPLE 7	4005	3985
196	554	551	SAMPLE 7	4005	3999
199	552	552	SAMPLE 7	4016	3995
202	553	553	SAMPLE 7	4020	4000
202	554	551	SAMPLE 7	4010	4005
203	552	550	SAMPLE 7	4000	4010
202	552	551	SAMPLE 12	4009	3993
199	552	551	SAMPLE 12	4012	3993
201	552	552	SAMPLE 12	4020	3998
196	553	551	SAMPLE 12	4016	4003
191	553	552	SAMPLE 12	4017	4002

(continued)

TABLE A2.3 Continued.

a	b	c	Sample	W	L
201	551	551	SAMPLE 12	4016	4002
201	552	551	SAMPLE 12	4012	3998
200	552	551	SAMPLE 12	4011	3995
203	550	551	SAMPLE 12	4003	3979
200	551	551	SAMPLE 12	4003	3979
201	552	548	SAMPLE 12	4008	3989
202	551	543	SAMPLE 12	4008	3987
201	550	550	SAMPLE 12	4012	3997

CASE STUDY 4.

The quality control in fruit juice is usually performed by comparing the values of several constituents in tested samples against standards derived from the analysis of a "base sample" composed of presumably pure products. Both the issues of the selection of the "base sample" and that of the proper methods of identification of possible adulteration have been debated in the professional literature. The choice of the "base sample" is critical to the quality control process. Previous studies have shown that the values of the attributes in fruits differ considerably between varieties. The values also vary due to environmental and climatic factors. Proposed "universal base samples" with the appropriate test procedures have been shown to reject far too many pure samples of natural juice.

We do not address here the the practical issues related to the selection of the "base sample". Rather we assume that a priori - before adulteration, if any took place - the "base sample" and the "test" samples come from the same population. This somewhat idealized case corresponds to a "base sample" created from the fruit juice used in the industrial process of juice extraction.

The data set in Case Study 4 is consistent with the assumption presented above. The data consist of the concentrations in miug per standard volume of 11 amino-acids (LYS=Lysine, ARG=Arginine, ASP=Aspartic acid, SER=Serine, GLU=Glutamine acid, PRO=Proline, GLY=Glycine, ALA=Alanine, VAL=Valine, PHA=Phenyl alanine, GABA=Gamma-amino butric acid) in 69 specimens of a pure juice.

The process of dilution and adulteration has been simulated as follows: a "base sample" of predermined size 36 has been randomly selected from the data set. For the remaining second sample of size 33 "tested" specimens, the measured values of the concentration variables were multiplied by α, representing a variables were "adulterated" by adding $100(1 - \alpha)\%$ of the means of those variables. This represents an "ideal adulteration" since the original means of these components were recovered (although not the original values).

The statistical methods are applied to the resulting samples and we assess their ability to detect the "adulteration". In the following data the "base sample" is followed by the "test" sample.

TABLE A2.4
The data from Case Study 4

LYS	ARG	ASP	SER	GLU	PRO	GLY	ALA	VAL	PHA	GABA
0.48	5.81	2.12	4.68	0.78	12.41	0.31	0.96	0.18	0.20	4.73
0.47	5.25	2.75	4.42	0.88	14.72	0.30	1.04	0.19	0.22	3.96
0.42	4.98	2.79	3.85	0.75	12.13	0.32	0.99	0.15	0.20	3.94
0.35	4.79	2.79	3.39	0.81	12.77	0.25	0.75	0.16	0.15	3.69
0.43	4.92	2.88	3.53	0.78	13.11	0.25	0.91	0.16	0.15	4.23
0.40	5.61	2.26	3.39	0.69	12.69	0.20	1.06	0.16	0.18	3.76
0.35	4.54	2.96	3.89	0.88	14.01	0.24	0.86	0.16	0.12	3.92
0.34	3.82	2.86	3.63	0.86	15.73	0.22	1.34	0.14	0.12	2.88
0.27	3.42	2.27	4.81	0.90	8.99	0.23	1.43	0.10	0.10	2.68
0.39	3.60	2.99	5.03	0.92	13.71	0.28	1.99	0.13	0.10	2.88
0.37	3.39	2.78	5.96	0.84	12.92	0.24	1.76	0.12	0.14	3.01
0.26	2.72	3.82	6.03	1.17	7.18	0.15	1.30	0.11	0.07	3.40
0.24	3.13	3.35	5.76	0.96	6.75	0.21	1.14	0.11	0.08	2.43
0.20	2.15	3.28	5.80	1.04	5.34	0.22	1.06	0.12	0.08	2.41
0.26	2.89	3.67	6.34	1.22	5.87	0.18	1.10	0.14	0.12	2.40
0.52	5.53	2.97	3.37	0.78	10.74	0.24	0.96	0.10	0.16	3.40
0.42	5.07	3.06	4.32	0.91	15.37	0.47	1.32	0.16	0.20	3.63
0.45	5.46	3.06	4.68	0.84	16.52	0.39	1.35	0.14	0.18	3.89
0.47	5.79	2.91	4.44	0.80	16.21	0.35	1.20	0.20	0.18	4.52
0.44	2.52	2.40	4.09	0.72	12.81	0.28	0.86	0.18	0.23	4.43
0.48	5.14	2.66	4.04	0.94	16.77	0.33	0.97	0.22	0.23	4.90
0.49	4.77	2.42	5.92	1.00	15.62	0.34	1.93	0.50	0.15	4.05
0.37	4.35	3.04	5.07	0.87	15.81	0.31	2.08	0.19	0.10	4.17
0.36	4.01	2.37	3.93	0.76	11.28	0.22	0.75	0.12	0.12	3.27
0.46	4.26	2.51	7.29	1.07	18.57	0.37	2.67	0.19	0.10	2.95
0.34	3.46	2.20	3.80	0.93	11.73	0.26	1.40	0.18	0.10	3.06
0.34	4.13	2.72	6.01	0.95	13.96	0.34	2.30	0.10	0.08	3.06
0.31	3.70	2.77	5.29	0.85	10.80	0.22	1.68	0.10	0.01	2.61
0.30	3.18	2.54	5.04	0.95	11.25	0.21	1.84	0.10	0.01	2.48
0.30	3.57	2.45	5.70	1.06	12.28	0.26	1.53	0.10	0.10	2.46
0.30	3.31	2.53	5.21	0.88	9.10	0.23	1.37	0.08	0.01	2.55
0.30	3.13	2.82	5.85	1.00	10.31	0.21	1.55	0.10	0.08	2.69
0.33	3.10	3.01	7.15	1.04	12.71	0.23	1.79	0.09	0.10	3.52
0.32	3.84	3.79	6.08	1.01	10.13	0.18	1.30	0.09	0.01	3.67

TABLE A2.4 Continued.

LYS	ARG	ASP	SER	GLU	PRO	GLY	ALA	VAL	PHA	GABA
0.30	3.75	2.83	6.24	0.71	6.20	0.16	1.20	0.05	0.08	3.01
0.26	3.34	3.46	7.01	1.02	6.68	0.20	1.52	0.10	0.08	2.18
0.43	5.84	2.84	3.54	0.80	11.90	0.30	0.86	0.20	0.18	3.88
0.50	4.61	2.08	5.70	0.71	18.46	0.42	1.91	0.18	0.18	6.14
0.51	6.19	3.55	4.29	1.16	19.01	0.40	1.20	0.15	0.18	4.72
0.43	5.44	2.71	4.38	0.79	13.59	0.35	1.23	0.14	0.20	4.08
0.38	5.22	2.54	3.97	0.73	14.47	0.30	0.98	0.15	0.20	4.18
0.50	5.19	3.13	4.32	0.90	16.74	0.34	1.09	0.20	0.22	4.85
0.40	4.68	2.38	3.47	0.68	12.01	0.26	0.92	0.16	0.18	3.95
0.43	4.99	2.03	3.52	0.63	9.84	0.24	0.71	0.19	0.20	4.06
0.41	5.33	2.64	4.22	0.81	13.66	0.30	0.86	0.17	0.22	4.52
0.45	5.42	2.96	4.80	0.91	15.73	0.30	1.09	0.19	0.20	3.80
0.36	4.83	2.72	3.32	0.75	12.28	0.23	0.71	0.13	0.12	3.63
0.40	4.34	1.92	4.57	0.74	11.13	0.28	1.63	0.14	0.10	3.34
0.36	4.41	2.88	3.76	0.89	14.32	0.25	0.89	0.14	0.12	3.35
0.30	4.14	2.50	5.26	0.86	15.48	0.35	2.34	0.21	0.12	3.02
0.38	3.91	2.32	5.14	0.82	14.27	0.29	1.87	0.22	0.10	3.98
0.42	3.90	2.45	5.26	0.94	18.14	0.29	2.03	0.16	0.12	3.65
0.31	3.56	2.61	5.40	0.97	12.29	0.22	1.59	0.08	0.10	2.82
0.32	4.18	3.76	5.53	0.98	10.81	0.22	1.32	0.10	0.14	2.91
0.32	3.05	3.24	6.87	1.43	13.01	0.24	1.81	0.10	0.10	2.91
0.23	3.13	3.43	6.30	1.15	10.67	0.26	1.67	0.12	0.16	2.86
0.24	2.85	3.18	4.64	0.86	6.91	0.21	1.08	0.01	0.12	2.75
0.36	4.31	2.25	3.15	0.65	11.32	0.22	0.83	0.19	0.20	3.66
0.35	4.62	2.40	2.94	0.71	10.18	0.19	0.89	0.19	0.20	3.01
0.39	4.51	2.82	4.00	0.87	13.76	0.27	0.88	0.17	0.12	3.56
0.41	4.12	2.38	5.14	0.83	11.36	0.26	1.71	0.16	0.08	3.65
0.33	3.60	2.36	5.07	0.94	13.93	0.30	1.62	0.10	0.08	3.51
0.43	4.11	2.22	6.86	1.12	14.35	0.27	1.68	0.10	0.12	3.96
0.31	3.70	2.77	5.44	1.02	12.68	0.32	1.75	0.10	0.10	3.47
0.36	3.64	2.21	6.56	1.02	15.53	0.39	1.96	0.10	0.10	3.07
0.27	3.25	2.82	4.92	0.91	8.43	0.20	1.53	0.10	0.01	2.32
0.28	2.91	3.21	6.41	1.35	9.42	0.22	1.80	0.12	0.01	2.85
0.30	3.64	2.73	5.76	0.73	5.55	0.20	0.94	0.05	0.05	3.14
0.28	2.68	3.61	6.38	1.06	6.94	0.22	1.22	0.11	0.11	2.71

Appendix 3

Review of Matrix Algebra for Statistics with MINITAB ™ Computations

We summarize in the present appendix some properties and techniques of matrix operations which are useful in statistical analysis. Basic results are presented but not proven. The interested reader can study the subject from one of the many textbooks on matrix algebra for statistics (such as Graybill, 1965). We use MINITAB to perform matrix calculations.

An array of n rows and 1 column is called an n-dimensional *vector*. We denote a vector with low-case bold type letter $\mathbf{a} = (a_1, a_2, \ldots, a_n)$ where a_i is the ith component of vector $\mathbf{a}$. The transpose of an n-dimensional vector $\mathbf{a}$ is an array of 1 row and n columns. We denote the transpose of $\mathbf{a}$ by $\mathbf{a}'$. We can add (subtract) only vectors of the same dimension. Thus $\mathbf{a} + \mathbf{b}$ is a vector $\mathbf{c}$ whose i-th component is $c_i = a_i + b_i$ $i = 1, \ldots, n$. The vector $\mathbf{0}$ is one whose components are all equal to zero. Obviously, $\mathbf{a} + (-\mathbf{a}) = \mathbf{0}$. If $\mathbf{a}$ and $\mathbf{b}$ are vectors of the same dimension, n, then the *inner product* of $\mathbf{a}$ and $\mathbf{b}$ is $\mathbf{a}'\mathbf{b} = \sum_{i=1}^{n} a_i b_i$. The Euclidean *norm* (length) of a vector $\mathbf{a}$ is $\|\mathbf{a}\| = (\mathbf{a}'\mathbf{a})^{1/2}$. If α is a scalar (real number) and $\mathbf{a}$ a vector then $\alpha\mathbf{a}$ is a vector whose i-th component is αa_i, $i = 1, \ldots, n$. If $\mathbf{a}_1, \mathbf{a}_2, \ldots, \mathbf{a}_k$ are n-dimensional vectors and $\alpha_1, \ldots, \alpha_k$ are scalars then $\sum_{i=1}^{k} \alpha_i \mathbf{a}_i$ is an n-dimensional vector.

A vector $\mathbf{b}$ is *linearly independent* of vectors $\mathbf{a}_i, \ldots, \mathbf{a}_k$ *if there exist no scalars* $\alpha_1, \ldots, \alpha_k$ *such that* $\mathbf{b} = \sum_{i=1}^{k} \alpha_i \mathbf{a}_i$ k vectors $\mathbf{a}_1, \ldots, \mathbf{a}_k$ are *mutually independent* if no one is a linear combination of the other $(k - 1)$ vectors.

A matrix of order $n \times k$ is an array of n rows and k columns, i.e., $\mathbf{A} = (\mathbf{a}_1, \mathbf{a}_2, \ldots, \mathbf{a}_k)$. An $n \times n$ matrix is called a squared matrix. An $n \times k$ matrix $\mathbf{A}$ will be denoted also as

$$\mathbf{A} = (a_{ij}, i = 1, \ldots, n, j = 1, \ldots, k) .$$

If $\mathbf{A}$ and $\mathbf{B}$ are matrices of the same order then $\mathbf{A} \pm \mathbf{B} = \mathbf{C}$, where

$$c_{ij} = a_{ij} \pm b_{ij}, \quad i = 1, \ldots, n, \quad j = 1, \ldots, k .$$

An $n \times n$ matrix $\mathbf{A}$, such that

$$a_{ij} = \begin{cases} a_i, & \text{if } i = j \\ 0, & \text{if } i \neq j \end{cases}$$

is called a **diagonal** matrix. The diagonal matrix $\mathbf{I}_n$, such that its diagonal elements are all equal to 1 and all the off-diagonal elements are equal to 0 is called the *identity* matrix of order n. The vector $\mathbb{1}_n$ will denote an n-dimensional vector of 1's. The matrix $\mathbf{J}_n$ will denote an $n \times n$ matrix whose elements are all equal to 1. Notice that $\mathbb{1}'_n \mathbb{1}_n = n$.

The product of $\mathbf{A}$ and $\mathbf{B}$ is defined if the number of columns of $\mathbf{A}$ is equal to the number of rows of $\mathbf{B}$. The product of an $(n \times k)$ matrix $\mathbf{A}$ by an $(k \times l)$ matrix $\mathbf{B}$ is an $(n \times l)$ matrix C such that

$$c_{ij} = \sum_{m=1}^{k} a_{im} b_{mj} ; \quad i = 1, \ldots, n, j = 1, \ldots, l .$$

Notice that two squared matrices of the same order can always be multiplied. Also

$$\mathbf{J}_n = \mathbb{1}_n \mathbb{1}'_n .$$

We illustrate these operations by using MINITAB.

In MINITAB, matrices are denoted by $M1$, $M2$, etc. Calculations with matrices can be found in the Menu, under 'Calc'. Matrices can be read from a text file, or from the keyboard. For example, to read in the matrix

$$\mathbf{A} = \begin{bmatrix} 13 & 7 & 25 \\ -4 & 15 & 5 \\ 6 & -2 & 20 \end{bmatrix}$$

we use the command

MTB> Read 3 3 $M1$.

DATA> 13 7 25

DATA> −4 15 5

DATA> 6 − 2 20

The data is entered one row at a time. One can also copy several columns of equal size into a matrix, e.g.

MTB > copy $C1 - C3$ $M5$.

The transpose of an $n \times k$ matrix $\mathbf{A}$, is a $k \times n$ matrix denoted by $\mathbf{A}'$ in which the i-th row in $\mathbf{A}$ is the i-th column in $\mathbf{A}'$. To obtain in MINITAB the transpose of $\mathbf{A}$ we write

MTB > transpose $M1$ $M2$.

$\mathbf{A}'$ resides in $M2$. We can now compute $\mathbf{B} = \mathbf{A}'\mathbf{A}$, by

MTB > multiply $M2$ $M1$ $M3$.

The matrix $\mathbf{B}$ is

$$\mathbf{B} = \begin{bmatrix} 221 & 19 & 425 \\ 19 & 278 & 210 \\ 425 & 210 & 1050 \end{bmatrix}$$

Notice that the matrix $\mathbf{B}$ is symmetric, i.e., $b_{ij} = b_{ji}$ for all $i \neq j$.

Suppose that $\mathbf{A}$ is an $n \times k$ matrix, and $k \leq n$. The rank of $\mathbf{A}$ is the maximal number of mutually independent column vectors of $\mathbf{A}$.

An $n \times n$ matrix of rank n is called *nonsingular*. Every nonsingular matrix $\mathbf{A}$ has an inverse $\mathbf{A}^{-1}$ satisfying

$$\mathbf{A}\mathbf{A}^{-1} = \mathbf{A}^{-1}\mathbf{A} = \mathbf{I}.$$

To obtain the inverse of a matrix using MINITAB we write

$$\text{MTB} > \text{inverse } M3 \ M4.$$

The inverse of the matrix $\mathbf{B}$ above is

$$\mathbf{B}^{-1} = \begin{bmatrix} 0.03277 & 0.00916 & -0.01510 \\ 0.00916 & 0.00680 & -0.00507 \\ -0.01510 & -0.00507 & 0.00808 \end{bmatrix}$$

The Hotelling's T^2 for a single multivariate observation $\underline{\mathbf{X}}' = (X_1, X_2, \ldots, X_p)$ relative to a target $\underline{\mathbf{m}}' = (m_1, m_2, \ldots, m_p)$ and with a covariance matrix $\mathbf{S}$ was defined as $T^2 = (\underline{\mathbf{X}} - \underline{\mathbf{m}})'\mathbf{S}^{-1}(\underline{\mathbf{X}} - \underline{\mathbf{m}})$. In order to compute it, we need to determine:

a) $(\underline{\mathbf{X}} - \underline{\mathbf{m}})$

b) $(\underline{\mathbf{X}} - \underline{\mathbf{m}})' = \text{transpose } (\underline{\mathbf{X}} - \underline{\mathbf{m}})$

c) $S^{-1} = \text{Inverse } (S)$

The MINITAB commands *transpose* and *inverse* can be used to perform these operations that produce the required multivariate distance T^2:

For example, let us compute T^2 for $\underline{\mathbf{X}}' = (23, 28, 35)$ $\underline{\mathbf{m}}' = (30, 30, 30)$ and

$$\underline{\mathbf{S}} = \begin{bmatrix} 25 & 20 & 60 \\ 20 & 30 & -15 \\ 60 & -15 & 20 \end{bmatrix}$$

1) MTB>READ 1 3 M11
   DATA> 23 45 15

2) MTB>READ 1 3 M12
   DATA>30 30 30

3) MTB>READ 3 3 M13
   DATA> 25 20 60
   DATA> 20 30 -5
   DATA> 60 -15 20

4) MTB>SUBTRACT M11 M12 M21

5) MTB>TRANSPOSE M21 M22

6) MTB>INVERSE M13 M23

7) MTB>MULTIPLY M22 M23 M24

8) MTB>MULTIPLY M24 M21 M25
   ANSWER=27.6473

A squared matrix $\mathbf{H}$ is called *orthogonal* if $\mathbf{H}^{-1} = \mathbf{H}'$. An example of orthogonal matrix is

$$\mathbf{H} = \begin{bmatrix} \frac{1}{\sqrt{3}} & -\frac{1}{\sqrt{2}} & \frac{1}{\sqrt{6}} \\ \frac{1}{\sqrt{3}} & 0 & -\frac{2}{\sqrt{6}} \\ \frac{1}{\sqrt{3}} & \frac{1}{\sqrt{2}} & \frac{1}{\sqrt{6}} \end{bmatrix}$$

Use MINITAB to verify that indeed $\mathbf{H}^{-1} = \mathbf{H}'$.

For every real symmetric matrix $\mathbf{A}$, there exists an orthogonal matrix $\mathbf{H}$ *such that*,

$$\mathbf{HAH}' = \Lambda ,$$

where Λ is a diagonal matrix. The elements $\lambda_{ii}\,(i = 1, \ldots, n)$ of Λ are called the *eigenvalues* of $\mathbf{A}$. Corresponding to the eigenvalue λ_{ii} there is an *eigenvector* $\mathbf{x}_i$, satisfying the linear equations

$$A_{\mathbf{x}_i} = \lambda_{ii}\mathbf{x}_i \quad (i = 1, \ldots, n) .$$

The eigenvalues and vectors of a symmetric matrix are obtained by the command

$$\text{MTB} > \text{eigen } M3 \, C1 \, M8 \,.$$

Here, the eigenvalues of $M3$ are stored in $C1$ and the eigenvectors are put in the matrix $M8$. The eigenvalues of **B** are thus

$$1269.98, \quad 255.73, \quad 23.29$$

and the matrix of eigenvectors is

$$\mathbf{P} = \begin{bmatrix} 0.371082 & 0.334716 & -9.866177 \\ 0.199177 & -0.939762 & -0.277847 \\ 0.907001 & 0.069367 & 0.415377 \end{bmatrix} \,.$$

Notice that **P** is an orthogonal matrix.

References

Alt, F. (1984), Multivariate Quality Control, in the Encyclopedia of Statistical Science, eds. S. Kots, N.L. Johnson, and C.R. Read, John Wiley and Sons, Inc., New York, NY.

Andrews, D.R. (1971), A Note of the Selection of Data Transformations, *Biometrika*, 58, pp. 249–254.

ASQC (1983), Glossary and Tables for Statistical Quality Control, American Society of Quality Control, Milwaukee, Wisconsin.

Becker, R., Cleveland, W., and Wilks, W. (1987), Dynamic Graphics for Data Analysis, *Statistical Science*, 2, pp. 355–395.

Box, G.E.P. and Cox, D.R. (1964), An Analysis of Transformations, *Journal of the Royal Statistical Society*, B, 26, pp. 211–252.

Box, G.E.P., Hunter, W.G. and Hunter, J.S. (1978), Statistics for Experimenters, John Wiley and Sons, New York, NY.

Chan, L.K., Cheng, S.W. and Spiring, F.A. (1991), A Multivariate Measure of Process Capability, *International Journal of Modelling and Simulation*, 11, pp. 1–6.

Chua, M.K., and Montgomery, D.C. (1991), A Multivariate Quality Control Scheme, *International Journal of Quality & Reliability Management*, 6, pp. 29–46.

Dempster, A.P. (1963), Multivariate Theory for General Stepwise Methods, *The Annals of Mathematical Statistics*, 34, pp. 873–883.

Eisenhart, C., Hastay, M. and Wallis, W. (1947), eds., Techniques of Statistical Analysis of Scientific and Industrial Research and Production and Management Engineering, McGraw-Hill, New York, NY.

Execustat 3.0®(1993), Student Edition, Duxbury Press, Belmont, California.

Fuchs, C. and Benjamini, Y. (1991), Multivariate Profile Charts of SPC, *Series in Applied Statistics Technical Report 91-01*. Department of Statistics, Tel Aviv University.

Fuchs, C. and Kenett, R.S. (1987), Multivariate Tolerance Regions and F-Tests, *Journal of Quality Technology*, 19, pp. 122–131.

Fuchs, C. and Kenett, R.S. (1988), Appraisal of Ceramic Substrates in Multivariate Tolerance Regions, *The Statistician*, 37, pp. 401–411.

Haslett, J., Bradley, R., Craig, P., Unwin, A. and Wills, G (1991), Dynamic Graphics for Exploratory Spatial Data with Application to Locating Global and Local Anomalies, *The American Statistician*, 3, pp. 234–242.

Hawkins, D.M. (1991), Multivariate Control Based on Regression-Adjusted Variables, *Technometrics*, 33, pp. 61–75.

Hawkins, D.M. (1993), Regression Adjustment for Variables in Multivariate Quality Control, *Journal of Quality Technology*, 26, pp. 197–208.

Holland, P. (1973), Covariance Stabilizing Transformations, *The Annals of Statistics*, 1, pp. 84–92.

Hotelling, H. (1931), The Generalization of Student's Ratio, *Annals of Mathematical Statistics*, 2, pp. 360–378.

Hotelling, H. (1947), Multivariate quality control illustrated by air testing of sample bombsights, Selected Techniques of Statistical Analysis, C. Eisenhart, et al., Editors, McGraw-Hill, New York.

Huber, P. (1987), Experiences with Three-Dimensional Scatter Plots, *Journal of the American Statistical Association*, 82, pp. 448–453.

ISO 3207-1975, Statistical Interpretation of Data-Determination of a Statistical Tolerance Interval, International Organization for Standardization, Geneva.

Jackson, J. (1959), Quality Control Methods for Several Related Variables, *Technometrics*, 1, pp. 359–377.

Jackson, J.E. (1985), Multivariate Quality Control, *Communications in Statistics, Theory and Methods*, 14, pp. 2656–2688.

JMP IN®(1989), Statistical Visualization Software for the Macintosh®, Duxbury Press, Belmont, California.

Juran, J.M. (1988), Quality Control Handbook, 4th edition, McGraw-Hill, New York.

Kabe, D.G. (1984), On the Maximal Invariance of MANOVA Step Down Procedure Statistics, *Communications in Statistics – Theory and Methods*, 13, pp. 2571–2581.

Kenett, R.S. and Halevy, A. (1984), Some Statistical Aspects of Quality Conformance Inspection in Military Specifications Documents, *Proceedings of the 5th International Conference of the Israeli Society for Quality Assurance*, Tel Aviv 6.2.1, pp. 1–10.

Kenett, R.S. and Zacks, S. (1998), Modern Industrial Statistics: Design and Control of Quality and Reliability, Duxbury Press, Belmont, California.

Koziol, J.A. (1993), Probability Plots for Assessing Multivariate Normality, *The Statistician*, 42, pp. 161–173.

Marden, J.I. and Perlman, M.D. (1980), Invariant Tests for Means and Covariates, *Annals of Statistics*, 8, pp. 25–63.

Marden, J.I. and Perlman, M.D. (1989), On the Inadmissibility of the Modified Step-Down Test Based on Fisher's Method for Combining p-Values, in Glaser et al. eds., Contributions to Probability and Statistics: Essays in Honor of Ingram Olkin, Springer, New-York, 472–485.

Marden, J.I. and Perlman, M.D. (1990), On the Inadmissibility of Step-Down Procedures for the Hotelling T^2 Problems, *Annals of Statistics*, 18, pp. 172–190.

Mardia, K.V. (1975), Assessment of Multinormality and the Robustness of Hotelling T^2 Test, *Applied Statistics*, 24, pp. 163–171.

Mosteller, F. and Tukey, J. (1977), Data Analysis and Regression, Addison-Wesley.

Mudholkar, G.S. and Subbaiah, P. (1980a), A Review of Step-Down Procedures for Multivariate Analysis of Variance, in Gupta, R.P. ed, Multivariate Statistical Analysis, North-Holland, Amsterdam.

Mudholkar, G.S. and Subbaiah, P. (1980b), Testing Significance of a Mean Vector – A Possible Alternative to Hotelling T^2, *Annals of the Institute of Mathematical Statistics*, 32, pp. 43–52.

Mudholkar, G.S. and Subbaiah, P. (1988), On a Fisherian Detour of the

Step-Down Procedure of MANOVA, *Communications in Statistics – Theory and Methods*, 17, pp. 599–611.

Murphy, B.J. (1987), Selecting Out of Control Variables with the T^2 Multivariate Quality Control Procedure, *The Statistician*, 36, 571–583.

Phadhe, M.S. (1989), Quality Engineering Using Robust Design, Prentice Hall, Englewood Cliffs, New Jersey.

Ryan, T.P. (1988), Statistical Methods for Quality Improvement, Wiley, New York.

Seber, G.A.F. (1984), Multivariate Observations, Wiley, New York.

Shewhart, S.A. (1931), Economic Control of Quality of Manufactured Product, D. Van Nostrand Company Inc., New York.

STATGRAPHICS 3.0® (1988), Statistical Graphics System, STSC, Rockville, Md.

Thisted, R. (1986), Computing Environments for Data Analysis, *Statistical Science*, 2, pp. 259–275.

Tukey, J. (1962), The Future of Data Analysis, *Annals of Mathematical Statistics*, 33, pp. 1–67. Corrected pp. 812.

Tukey, J. (1977), Exploratory Data Analysis, Addison-Wesley.

Tukey, J. (1986), Data-Based Graphics: Visual Display in the Decades to Come, *Statistical Science*, 5, pp. 327–339.

Velleman, P. (1989), Learning Data Analysis with Data Desk®, W.H. Freeman, New York.

Velleman, P. and Hoaglin, D. (1981), Basics and Computing of Exploratory Data Analysis, Duxbury Press.

Wadsworth, H.M., Stephens, K.S. and Godfrey, A.B. (1986), Modern Methods for Quality Control and Improvement, J. Wiley, New York.

Weihs, C. and Schmidt, H. (1990), OMEGA (Online Multivariate Exploratory Graphical Analysis): Routing Searching for Structure – with Discussion, *Statistical Science*, 2, pp. 175–226.

Western Electric (1956), Statistical Quality Control Handbook, Western Electric Corporation, Indianapolis, Indiana.

Wierda, S.J. (1993), A multivariate Process Capability Index, ASQL Quality Congress Transactions, Boston.

Wierda, S.J. (1994), Multivariate Statistical Process Control, Wolters-Noordhoff, Groningen.

Index

Printed and bound by CPI Group (UK) Ltd, Croydon, CR0 4YY

24/10/2024

01778277-0002